It Worked for Me

ALSO BY COLIN POWELL

My American Journey

It Worked for Me

In Life and Leadership

Colin Powell

WITH TONY KOLTZ

HarperCollins books may be purchased for educational, business, or sales promotional use. For information please write: Special Markets Department, HarperCollins Publishers, 10 East 53rd Street, New York, NY 10022.

FIRST HARPERLUXE EDITION

Library of Congress Cataloging-in-Publication Data is available upon request.

ISBN: 978-0-06-218406-1

12 13 14 ID/RRD 10 9 8 7 6 5 4 3 2 1

To

Jeffrey, Bryan, Abby, and PJ

Our future

Contents

It Worked for Me

Author's Note

I love stories. In the course of my career, I gathered a number of them that mean a lot to me. Most come from my military life. I was in the military from age seventeen, as an ROTC cadet, until I was a retired GI at age fifty-six. Others came from my service as Secretary of State or National Security Advisor. Yet others came to me as I just wandered through life. In this book I want to share with you a selection of these stories and experiences that have stayed with me over the years; each one of them taught me something important about life and leadership. I offer them to you for whatever use you may wish to make of them.

Part I explains my "Thirteen Rules," which have been bouncing around since they were first published in *Parade* magazine over twenty years ago. Part II focuses on the importance of really knowing who you are and how to always be yourself. The emphasis in Part III is on knowing and taking care of others, especially those who are your followers. Part IV captures my experience in the exploding digital realm that has reshaped the world and our lives. Part V deals with how to be a great manager and a great leader. Part VI, "Reflections," describes serious and amusing aspects of my life. The afterword summarizes what the whole book is about—people in all their glorious, loving, and frustrating forms.

As you will see, there are no conclusions or recommendations, just my observations. The chapters are freestanding. You can read them straight through or jump in anywhere. Everyone has life lessons and stories. These are mine. All I can say is that they worked for me.

—Colin Powell

It Worked for Me

PART I
The Rules

Chapter One
My Thirteen Rules

President George H. W. Bush was sworn in to succeed President Ronald Reagan on January 20, 1989. The moment he took the oath I ceased to be the National Security Advisor; the torch was passed to my longtime colleague and mentor, General Brent Scowcroft.

After I left the White House, I returned to the Army. In April I was promoted to four-star general and given command of the Army's Forces Command (FORSCOM), with headquarters at Fort McPherson, Georgia, just outside Atlanta. I had command of all the deployable Army forces in the United States, including the Army Reserve, and I supervised the training of the Army National Guard. I was the first black Army officer to have a four-star troop command.

Shortly after I arrived at FORSCOM, *Parade* magazine, the long-running Sunday supplement with a readership of more than fifty million people, asked to do a cover story about me and my new assignment—one of those short personal articles aimed at Americans reading their Sunday newspapers over coffee. Since the story was written and the supplement printed many weeks before its August 13 distribution date, *Parade* had no way of knowing that the 13th would be just three days after I was announced by President Bush to be the next Chairman of the Joint Chiefs of Staff. The article was so timely that I was not able to persuade everyone that its publication date was a coincidence.

Its author, David Wallechinsky, a highly skilled journalist, needed a hook to close the piece. One of my secretaries, Sergeant Cammie Brown, urged him to ask me about the couple of dozen snippets of paper shoved under the glass cover on my desktop—quotes and aphorisms that I had collected or made up over the years. David called and asked if I would read off a few. The thirteen I read him appeared in a sidebar in the article.

After they were first printed in *Parade*—to my great surprise—the Thirteen Rules caught on. Over the past twenty-three years, my assistants have given out hundreds of copies of that list in many different forms; they have been PowerPointed and flashed around the world on the Internet.

Here are my rules and the reasons I have hung on to them.

1. IT AIN'T AS BAD AS YOU THINK. IT WILL LOOK BETTER IN THE MORNING.

Well, maybe it will, maybe it won't. This rule reflects an attitude and not a prediction. I have always tried to keep my confidence and optimism up, no matter how difficult the situation. A good night's rest and the passage of just eight hours will usually reduce the infection. Leaving the office at night with a winning attitude affects more than you alone; it also conveys that attitude to your followers. It strengthens their resolve to believe we can solve any problem.

At the Infantry School, they drilled into us constantly that an infantry officer can do anything. "No challenge is too great for us, no difficulty we cannot overcome." Think back to Churchill telling the world that Britain will "never, never, never give up." Or more colloquially, "Don't let the bastards get you down."

"Things will get better. You will make them get better." We graduated believing that, and I continue to believe that, despite frequent evidence to the contrary.

A variation of this theme was also drilled into us: "Lieutenant, you may be starving, but you must never show hunger; you always eat last. You may be freezing or near heat exhaustion, but you must never show that you are cold or hot. You may be terrified, but you must never show fear. You are the leader and the troops will reflect your emotions." They must believe that no matter how bad things look, you can make them better.

I love old movies and get from them lots of examples that I use for personal reinforcement.

The classic movie *The Hustler* opens with one of my all-time favorite scenes. It's set in a New York pool hall. A young pool whiz, Eddie Felson, played by Paul Newman, has come to challenge the reigning master, Minnesota Fats, played by Jackie Gleason. Also present are the pool impresario, Bert Gordon, a Mephistophelean figure played by George C. Scott, and a handful of spectators.

The match begins, and it is clear that Fast Eddie Felson is very good—maybe great. He proceeds to get the edge on Minnesota Fats, game after game, long into the evening. Fats starts to sweat. Others gather around to watch. Fast Eddie and his manager begin to smell triumph. The king is about to die; long live the new king. Fats, ready to give up, looks over to Bert for relief from the misery. Bert simply says, "Stay with this kid, he's a loser." Bert is a gambler and detects a weakness in Fast Eddie, an overconfidence that can be taken advantage of. Fats still seems stricken. He excuses himself and goes into the restroom. After washing his hands and face he comes out, seeming ready to leave. He signals to the attendant, and Fast Eddie smiles in victory, thinking Fats is asking for his coat. But no, Fats extends his hands for the attendant to apply talcum powder. Then, with a catlike smile he says, "Fast Eddie, let's play some pool." You know the rest—he crushes Eddie.

Many times when facing a tough meeting, an unpleasant encounter, a hostile press conference, or a vicious congressional hearing, the last thing I would do beforehand was go into the restroom, wash and dry my hands and face, look into the mirror, and say softly to myself, "Fast Eddie, let's play some pool." I may be down, but never out. An infantry officer can do anything.

Oh, full disclosure: Paul Newman is the star. At the end of the movie there is a rematch and he beats Fats. I never watch that scene.

2. GET MAD, THEN GET OVER IT.

Everyone gets mad. It is a natural and healthy emotion. You get mad at your kids, your spouse, your best friends, your opponents. My experience is that staying mad isn't useful. That experience was tested by my colleague the French foreign minister Dominique de Villepin, who made me—and most Americans—mad, very mad.

Dominique was a career diplomat, a graduate of the most prestigious French national academies, a noted historian, and a gifted poet, and he was very close to the then president of France, Jacques Chirac. With his flowing silvering hair and impeccable suits and ties, he cut quite a figure.

In early 2003, the period leading up to the Second Gulf War, there were repeated debates on that issue in the United Nations Security Council. The presidency of the fifteen-member council rotates every month; France had the presidency in January, with Dominique in the chair. The French were strongly opposed to military action against Iraq and led the opposition to it. They were not alone in their opposition. Germany, Russia, and a number of other countries had joined them. It's likely that more countries opposed us than supported us.

Council presidents normally suggest a special topic of discussion during their tenure. The topic Dominique suggested for a meeting of the fifteen Security Council foreign ministers was terrorism.

I was uneasy about this meeting. Would it stay on target? Most of my colleagues back in Washington thought the French would convert it into a session about Iraq—a bad idea; they wanted Iraq off the table at the UN. Dominique assured me, however, that it would stay focused on terrorism; there would be no discussion of Iraq. I accepted his assurances.

The meeting turned out fine . . . until Dominique left the conference to speak to the large assembled press corps, where he attacked our position on Iraq and made it clear that France would oppose any movement toward military action. I was blindsided; the White House phones lit up. The TV evening news and the press the next day made my embarrassment complete. The press loved the story, but it made life very difficult for me in Washington and at the UN. I was livid and made that clear to Dominique. Meanwhile, the reaction around the country was outrage. Newspapers were calling for a boycott of French wine and for renaming French fries freedom fries. In a nutshell, Dominique had screwed things up for me.

Dominique was in no way a bad man. He was reflecting the position of his government, he would remain the French foreign minister, and he came out looking like a hero to those who opposed us. For several months, Dominique would be my adversary on the Iraq issue, but I knew I could not treat him as an enemy.

Despite the opposition at the UN and elsewhere, President George W. Bush decided on military action and we deposed Saddam Hussein.

In the aftermath of the fall of Hussein, when we needed UN resolutions to restore order and rebuild Iraq, France supported us for six straight UN resolutions.

In February 2004, a crisis in Haiti required us to encourage President Jean-Bertrand Aristide to step down from office and leave the country. As mobs approached his home, we were able to get President Aristide and his party to the airport and on a U.S. plane heading to South Africa, where he thought he would be welcomed. That was a mistake. South Africa refused to receive him at that time. In the middle of the night I called Dominique to ask him to persuade one of the Francophone African countries to accept Aristide before our plane ran out of gas. Half an hour later he called back with a solution, and our anxious pilot soon had clear instructions on where to deposit President Aristide. My colleague and friend had come to my rescue.

We then sent a force in to stabilize Haiti until a UN force could be assembled, with a U.S. Marine general in charge. He had under his command a French infantry battalion. Dominique made that happen. These actions were in France's interest, but he could have made our lives a lot harder if I had made him an eternal enemy rather than an ally and friend who was an occasionally annoying adversary. I often remind folks that France was with us during the American Revolution. We have been married to the French for more than 230 years . . . and in marriage counseling with them for more than 230 years; but the marriage is still intact thanks to our shared values and common belief in human rights, freedom, and democracy. The ties that bind us are stronger than the occasional stresses that separate us.

Years ago, as a brigadier general stationed at Fort Leavenworth, Kansas, I worked for a great soldier, Lieutenant General Jack Merritt. I was in charge of evaluating how the Army should be organized and equipped in the future. General Merritt and I got along well, but one day he made a decision that I thought was shortsighted, unfair, and totally wrong. I asked to see him. When I went in and unloaded on him, he listened patiently with no visible emotion. After I finished my diatribe, he came over to me, put his hand on my shoulder, and quietly said, "Colin, the best part about being mad and disappointed is that you get over it. Now have a nice day." He was right. I felt better after getting my anger out, and I did get over it.

Jack Merritt was not the first to teach me this lesson. I originally learned it many years earlier in Germany as a young first lieutenant and company executive officer. One day I got into a screaming match on a phone with another officer and pretty much lost it. My commander, Captain William Louisell, observed my behavior. When I hung up, he said to me: "Don't ever act that way in my presence or anyone's presence again." To make sure I'd learned the lesson he wrote in my efficiency report, "Young Powell has a severe temper, which he makes a mature effort to control." He nailed me, but also gave me a life preserver. I've worked hard over the years to make sure that when I get mad, I get over it quickly and *never* lose control of myself. With a few lapses I won't discuss here, I've done reasonably well.

3. AVOID HAVING YOUR EGO SO CLOSE TO YOUR POSITION THAT WHEN YOUR POSITION FALLS, YOUR EGO GOES WITH IT.

I got this one from a couple of lawyers. Back in 1978, working as a staff assistant to Secretary of Defense Harold Brown during the Carter administration, I had to referee a heated dispute over some obscure issue. I sat at the head of the table in Secretary Brown's conference room filled with people and listened to two lawyers go at each other. They quickly got past the merits and demerits of the issue, but the debate continued, and for one of the lawyers it became increasingly personal. As he grew more and more agitated, he got himself tied up in arguments about how the outcome would affect him. I finally lost patience and stopped the debate. I'd heard enough. I decided the issue in favor of the other lawyer, based on the strength of his presentation and reasoning.

The fellow who lost looked crushed, to the discomfort of everyone in the room. The other lawyer looked at him and said, "Never let your ego get so close to your position that when your position falls, your ego goes with it." In short, accept that your position was faulty, not your ego.

This doesn't mean you don't argue with passion and intensity. During Secretary Brown's tenure, W. Graham Claytor was Deputy Secretary of Defense, and I was his military assistant. Graham was a crusty old Virginian, tough as nails, with high-level executive experience in government and the private sector. Before becoming deputy secretary, he had been Secretary of the Navy, and in private life he had been a distinguished lawyer, president of Southern Railway, and head of Amtrak. I've watched Graham go head to head with everyone in sight to advocate a position. If he lost the argument, he became a no less passionate advocate for what Secretary Brown had decided.

I encouraged all my subordinate commanders and staff to feel free to argue with me. My guidance was simple: "Disagree with me, do it with feeling, try to convince me you are right and I am about to go down the wrong path. You owe that to me; that's why you are here. But don't be intimidated when I argue back. A moment will come when I have heard enough and I make a decision. At that very instant, I expect all of you to execute my decision as if it were your idea. Don't damn the decision with faint praise, don't mumble under your breath—we now all move out together to get the job done. And don't argue with me anymore unless you have new information or I realize I goofed and come back to you. Loyalty is disagreeing strongly, and loyalty is executing faithfully. The decision is not about you or your ego; it is about gathering all the information, analyzing it, and trying to get the right answer. I still love you, so get mad and get over it."

No one followed this guidance better than Marine Colonel Paul "Vinny" Kelly, my congressional affairs assistant when I was Chairman of the Joint Chiefs of Staff. Vinny's job was to get me up on Capitol Hill as often as he could to testify, chat with members, hustle staff, and do all the other things that put you on the right side of the folks who allocate the people's money. I understood the importance of this activity, but Vinny was always pressing me to do more. He would come into my office late in the evening, after a trying day, to press me to attend another congressional meeting I didn't think was necessary. We would get into all kinds of arguments, which usually ended with "Vinny, get the hell out of here!" He would leave, disappointed, but accepting. The next day he would be back with new reasons why I had to go up to the Hill. These usually won me over. Vinny knew that "get the hell out" was not about him. His ego was never on his sleeve. He accepted my decision; yet he also knew that his job was to protect me, and so if he still thought he was right and I was wrong, he marshaled new arguments. He also knew Rule 1, "It will look better in the morning." He was a treasure. When I became Secretary of State, I pulled him off his retirement golf course and made him my Assistant Secretary of State for Legislative Affairs.

4. IT CAN BE DONE.

This familiar quotation is on a desk plaque given to me by the great humorist Art Buchwald. Once again, it is more about attitude than reality. Maybe it can't be done, but always start out believing you can get it done until facts and analysis pile up against it. Have a positive and enthusiastic approach to every task. Don't surround yourself with instant skeptics. At the same time, don't shut out skeptics and colleagues who give you solid counterviews. "It can be done" should not metamorphose into a blindly can-do approach, which leaves you running into brick walls. I try to be an optimist, but I try not to be stupid.

5. BE CAREFUL WHAT YOU CHOOSE: YOU MAY GET IT.

Nothing original in this one. Don't rush into things. Yes, there are occasions when time and circumstances force you to make fast decisions. Usually there is time to examine the choices, turn them over, look at them in the light of day and the darkness of night, and think through the consequences. You will have to live with your choices. Some bad choices can be corrected. Some you'll be stuck with.

6. DON'T LET ADVERSE FACTS STAND IN THE WAY OF A GOOD DECISION.

Superior leadership is often a matter of superb instinct. When faced with a tough decision, use the time available to gather information that will inform your instinct.

Learn all you can about the situation, your opponent, your assets and liabilities, your strengths and weaknesses, the threats and risks. Select several possible courses of action, then test the information you have gathered against them and analyze one against the other. Often, the factual analysis alone will indicate the right choice. More often, your judgment will be needed to select from the best courses of action. This is the moment when you apply your instinct to smell the right answer. This is where you apply your education, experience, and knowledge of external considerations unfamiliar to your staff. This is when you look deep into your own fears, anxiety, and self-confidence. This is where you earn your pay and position. Your instinct at this point is not a wild guess or a hunch. It is an informed instinct that knows from long experience which facts are the most important and which adverse facts, however adverse, can be set aside. As the saying goes, "Good judgment comes from experience, and experience comes from bad judgment."

On the eve of D-Day, General Eisenhower faced one of the most difficult decisions any military commander has ever had to make. The weather was dicey; launching the invasion into bad weather could doom it, but his weathermen predicted a possible opening on June 6, 1944. He had been gathering information and planning this operation for months. He knew it in his fingertips.

In the loneliness that only commanders know, he made his decision. He wrote a statement taking all the blame if the invasion failed. Yet his informed instinct said, "Go!" He was right.

In the final weeks of the Civil War, General Grant's Army of the Potomac was besieging Petersburg and slowly squeezing General Lee's Army of Northern Virginia to death. One night Grant was awakened by a staff officer. "We've received information that Lee's army is on the move and massing to attack our flank," he told Grant urgently. Grant rubbed the sleep from his eyes, thought for a moment, and said, "That's not possible," and went back to sleep.

Both generals could have been wrong, and history would have treated them differently. Eisenhower was a masterful staff officer and a gifted manager, but also a great leader. He knew when to trust his instinct. Grant did not make a snap judgment that night. He knew Lee, he had studied him as a man and soldier, and he knew the strengths and increasing weakness of the Army of Northern Virginia. His instinct was well informed, and it took only a minute for his instinct to conclude, "That's not possible."

There will be times when an adverse fact should stop you in your tracks. Never let it stop you completely until you have thought about it, challenged it, and looked for a way to get around it. And if you conclude that the gain will be great enough to overcome the consequences of that adverse fact, decide and execute.

I dare not compare myself to Eisenhower or Grant, but a similar though far smaller decision came my way in December 1989, a few months after I became JCS Chairman. On the night of December 1 there was an attempted military coup in the Philippines against President Corazon Aquino. I raced down to the command center in the Pentagon to monitor the action. President Aquino was concerned that members of the air force would join the coup and bomb the presidential palace. She called the White House and asked us to bomb the nearby air base to keep that from happening. I got instructions from the White House situation room to execute the mission. My experience told me it was an easy mission using F-4 Phantom jets from Clark Air Base. My experience also told me that there would be Filipino deaths and collateral damage to property. Regardless of how the coup turned out, Filipinos would surely criticize us for any loss of life and property damage. My instinct told me there might be a better way to accomplish the goal of the mission, which was to keep the palace from being bombed. Admiral Hunt Hardisty, our commander in the Pacific, happened to be in Washington and joined me at the command center. The alternative we came up with was to instruct the F-4 pilots to take off and buzz the Philippine air base in a manner that demonstrated "extreme hostile intent." If a plane took to the runway anyway, shoot in front of it or crater the runway. If the plane took off, then shoot it down. The Philippine planes stayed on the ground, and the coup ended a few hours later.

If one plane had managed to get off, bomb the palace, and kill the president, my experience and instincts would have failed.

During the crisis, I wasn't able to reach the Philippine minister of defense, Fidel Ramos. After it was all over I finally got through to him and briefed him on what we had done. He was deeply grateful that we had not bombed.

Whenever I'm faced with a difficult choice, my approach has always been to make an estimate of the situation—a familiar military process: What's the situation? What's the mission? What are the different courses of action? How do they compare with one another? Which looks most likely to succeed? Now, follow your informed instinct, decide, and execute forcefully; throw the mass of your forces and energy behind the choice. Then take a deep breath and hope it works, remembering that "hope is a bad supper, but makes a good breakfast."

7. YOU CAN'T MAKE SOMEONE ELSE'S CHOICES. YOU SHOULDN'T LET SOMEONE ELSE MAKE YOURS.

We are taught in the military to take full responsibility for "everything your unit does or fails to do, and what you do or fail to do." Since ultimate responsibility is yours, make sure the choice is yours and you are not responding to the pressure and desire of others.

That does not mean your decision has to be solitary or lonely. Seek the advice of others, but be aware that people are always around who are full of advice and sure they know how you should decide. All too often, your decision affects them and they are pushing you in a direction that's more in their interest than yours. Never forget that your informed instinct is usually the most solid basis for making a decision.

Of course, the choice is not always yours to make. In the Army, for instance, duty will at times require acceptance of that reality.

In 1985, I was selected to be an infantry division commander in Germany. I wanted the job badly—it is the dream job of every infantry officer, and I was eager to get back to troop command. But the Army decided I should remain in the Pentagon, continuing to serve as the senior military assistant to Secretary of Defense Caspar Weinberger.

A year later, I was able to leave the Pentagon and take command of a corps in Germany, an even larger unit. I was elated, but after six months, I was called back to Washington to serve as Deputy National Security Advisor. Since it seemed that would end my military career, I resisted. If it was that important, I asked, shouldn't the President call me? He did, and I left my corps. Eleven months later I became National Security Advisor for the remainder of President Reagan's term.

It's hard to fault the choices the Army made for me. Most of them turned out to be superb. But I have had more freedom to follow my own instincts and choices since I left the Army.

It's easy to be flattered into a job. When I left the State Department, I was flattered by offers of top positions in major corporations, most of them in the financial world. The monetary rewards were stunning and the work not terribly demanding. I was told I didn't need to know anything about banking, finance, or exotic financial instruments like hedge funds and derivatives. Experts would be present to help me. One investment bank pressed me hard, repeatedly upping the money and the title. The offers were definitely tempting.

I understood the financial and social value of these positions. But my instincts said no. Did they want me for what I could do for them? Or did they want me for the celebrity I could bring them? My instincts said I would mostly be a door opener and a dinner host. And the truth was I didn't have any relevant experience or background in the business, nor any desire to learn it. I couldn't care less about finance. In the end, I preferred my flexibility and independence. They were trying hard to make a choice for me, but I held out for my own choice.

One of my best friends helped me shape my instinct. Over lunch, he listened as I laid out all the offers. He replied simply, "Why would you want to wear someone else's T-shirt? You are your own brand. Remain free and wear your own T-shirt."

As it turned out, my instinct turned out to be not only right about my immediate choice, but also prescient. Most of the promised monetary rewards I passed up turned out to be fairy tale money. Firms that offered me top jobs either failed or came close to failure in the 2008 crash and ensuing recession. I'm glad I dodged that bullet!

These temptations pale in comparison with the choice I faced in 1995, two years after retiring from the Army. In those two years, I stayed out of the public eye, enjoyed private life, wrote my memoirs, and traveled the country speaking. But when my book was published and I went on a six-week book tour, I became more public than ever. The crowds were overwhelming. I had never imagined I'd get that kind of turnout. At every appearance the issue of running for political office arose. People were talking about me as a presidential candidate. It was incredibly flattering.

Though I've never had political ambitions, all the attention forced me to consider running. I debated what to do. What was best for me, my family, the nation? I reached out to friends and experts, and listened carefully to new friends who pushed me to run. A strong instinct told me that I had an obligation, a duty, to run. I had ideas about where the country should go and about how to fix what I saw was broken. But I was divided. An equally strong instinct warned that running for president would be a terrible choice for me.

The two months when I wrestled with that decision were perhaps the most difficult of my life. I was deeply conflicted, lost weight, had trouble sleeping. My family was split, which didn't make my choice easier. My very closest friends argued against running but were willing to help if I decided to choose that course. They knew me as well as I knew myself and felt a presidential campaign was not right for me.

The decision was mine to make. What drove my final choice was the reality that I did not wake up a single morning wanting to be president or with the fire and passion needed for a successful campaign. I was not a political figure. It was not me. Once I accepted what that instinct was telling me, the choice was clear, the decision easy.

I get asked almost daily if I have any regrets. The answer is no. It was my choice, my family's choice, and the right choice. I have no regrets and no reason to second-guess. I moved on and found other things to do to satisfy my need and my responsibility to serve the country. I disappointed many people but left others happy. It was my choice. It had to be.

8. CHECK SMALL THINGS.

We are all familiar with the old rhyme that begins, "For want of a nail . . ." It reminds us how small actions can result in large consequences.

Success ultimately rests on small things, lots of small things. Leaders have to have a feel for small things—a feel for what is going on in the depths of an organization where small things reside. The more senior you become, the more you are insulated by pomp and staff, and the harder and more necessary it becomes to know what is going on six floors down.

One way is to leave the top floor and its grand accoutrements and get down into the bowels for real. Don't tell anyone you are coming. Avoid advance notices that produce crash cleanups, frantic preparations, and PowerPoint presentations. Yes, sometimes you need to give lots of notice so folks can prepare their homes as if they were selling them. But I always preferred to just drop in and wander around. A maintenance shop with dirty mechanics, parts strewn around, and no senior officers lurking told me more about the state of maintenance than any formal quarterly reports.

Whenever I inspected barracks, I looked over the bunks and the displays of wall lockers and footlockers (long gone; troops now live in barracks resembling small college dormitories). I also made a beeline for the latrine. Not just to see if it was clean. Was there a shortage of toilet paper, were any mirrors cracked, were there any missing showerheads? Finding any of these situations immediately told me one of several things—the unit is running short of upkeep money, no one is checking on these things to get them fixed, or the troops are not being supervised well enough. Find out which and fix it.

I detested whitewashed rocks lining a pathway. And the smell of fresh paint meant they'd heard I was coming. Fresh cookies were another dead giveaway.

Once in Korea, we got word that the admiral commanding Pacific forces would be visiting our post and would walk through my battalion area. I was delighted. We lived in ancient, disgusting Quonset huts; we couldn't get parts for the stoves or paint for the outside. Because we were short of paint, I was told to paint the front but not the back of the mess hall the admiral would walk by. He walked by and saw the fresh paint. It was so fresh compared to everything else he saw that he wasn't fooled. We should have sat down and told him our problems and not forced him to be a detective.

The followers, the troops, live in a world of small things. Leaders must find ways, formal and informal, to get visibility into that world. In addition to my dropping in, I relied on a cast of informal observers who had direct access to me to tell me about details the system would not normally offer up to me. They also told me when I was totally screwed up and the "commander had no clothes." In my military commands, they were my chaplains, my command sergeant major and his network, my inspector general, and GIs coming in on "Open Door" night. In my National Security Council and State positions, I always had trusted friends outside and agents inside the organization who prowled the basement and kept me informed. Leaders need to know ground truth and not just what they get from reports and staffs.

One day at the State Department, about two in the afternoon, I was wandering around and ran into a young lady leaving the building. She did not seem to recognize me, or else she didn't let me know that she recognized me. I asked her why she was leaving so early. "I'm on flextime," she told me. "I started at seven a.m."

That got me curious; I didn't know much about flextime. I fell in stride with her and talked about how it worked for her and her fellow employees. I learned more about the program than I had ever heard from my staff. It was a good program, I realized—worth expanding. Meanwhile, she still didn't acknowledge who I was.

To needle her, I said, "Gee, I'd like to get flextime. How did you do it?"

"Ask your immediate supervisor," she responded.

"I'll do that on Monday, after he comes down from Camp David," I told her.

She didn't miss a beat. "Good," she said. "I hope you get it." She went through the door and I stood there not knowing if I'd been had. But I had learned a lot about flextime, a small thing for me, but a big thing for her and lots of my employees.

9. SHARE CREDIT.

When something goes well, make sure you share the credit down and around the whole organization. Let all employees believe they were the ones who did it. They were. Send out awards, phone calls, notes, letters, pats on the back, smiles, promotions—anything to spread the credit. People need recognition and a sense of worth as much as they need food and water.

In the military we make a big deal of change-of-command ceremonies, where the new commander assumes responsibility of the unit from the old commander, symbolized by the passing of the unit colors. These ceremonies normally function as celebrations of the commanders. The troops are assembled in formation on the parade field. The dignitaries arrive and the old and new commanders make speeches. The old commander is praised and given an award. The troops stand and listen, usually in the sun.

Lieutenant General Hank "the Gunfighter" Emerson, one of our most colorful generals and one of my favorite commanders, was not fond of these ceremonies. When I took command of my battalion in Camp Casey, Korea, he was my division commander. At that change-of-command ceremony, at his insistence, only the two commanders, their staff, and the company commanders stood in the middle of the field. No troops stood behind them, but they were invited to sit in the bleachers and watch the two senior officers pass the battalion colors from the old to the new commander. There were no speeches. I loved it.

A few years later, it became time for the Gunfighter to give up command of the XVIII Airborne Corps at Fort Bragg, North Carolina, home of the famous 82nd Airborne Division. Protocol and expectation required the old-fashioned ceremony with thousands of troops. I was then a brigade commander in the 101st Airborne Division at Fort Campbell, Kentucky, which was part of his corps. He ordered me to Fort Bragg to command the formation at his change-of-command and retirement ceremony.

After we'd practiced the ceremony to perfection, the day came. As we stood there in the sun waiting for it to begin, the Gunfighter signaled me to come up to the reviewing stand for new instructions. He directed me to return to the formation and order all the officers to do an about-face and gaze at their troops. I was then to order the officers to salute their soldiers. We conducted the ceremony, and the officers turned as he had directed and saluted the troops. It was a deeply moving moment. The gesture was the only way he could truly show that credit for his success belonged to the soldiers who had served under him.

It is the human gesture that counts. Yes, medals, stock options, promotions, bonuses, and pay raises are fine. But to really reach people, you need to touch them. A kind word, a pat on the back, a "well done," provided one-on-one and not by mob email is the way you share credit. It is the way you appeal to the dreams, aspirations, anxieties, and fears of your followers. They want to be the best they can be; a good leader lets them know it when they are.

When things go badly, it is your fault, not theirs. You are responsible. Analyze how it happened, make the necessary fixes, and move on. No mass punishment or floggings. Fire people if you need to, train harder, insist on a higher level of performance, give halftime rants if that shakes a group up. But never forget that failure is your responsibility.

Share the credit, take the blame, and quietly find out and fix things that went wrong. A psychotherapist who owned a school for severely troubled kids had a rule: "Whenever you place the cause of one of your actions outside yourself, it's an excuse and not a reason." This rule works for everybody, but it works especially for leaders.

10. REMAIN CALM. BE KIND.

Few people make sound or sustainable decisions in an atmosphere of chaos. The more serious the situation, usually accompanied by a deadline, the more likely everyone will get excited and bounce around like water on a hot skillet. At those times I try to establish a calm zone but retain a sense of urgency. Calmness protects order, ensures that we consider all the possibilities, restores order when it breaks down, and keeps people from shouting over each other.

You are in a storm. The captain must steady the ship, watch all the gauges, listen to all the department heads, and steer through it. If the leader loses his head, confidence in him will be lost and the glue that holds the team together will start to give way. So assess the situation, move fast, be decisive, but remain calm and never let them see you sweat.

The calm zone is part of an emotional spectrum that I work to maintain.

I try to have, and every leader should try to have, a healthy zone of emotions. Within that zone you can be a little annoyed, a little mad, a little loving. Within your zone you are calm (most of the time). You are interested. You are caring, yet you maintain a reasonable distance. You are consistent and mostly predictable (which does not mean you are dull and boring, or that you will never surprise them, or that you will never explode and come down hard on somebody). Your staff knows pretty much what your zone is and how to act accordingly.

Sometimes I *do* explode. Sometimes my explosions are right and justified.

One day, back in the hard-drinking old days in the Army, I was at wit's end dealing with DUI incidents. I was a brigade commander. A sergeant was standing before me about to be punished for driving under the influence. It was a serious offense. He knew he was facing a reduction in rank and a fine. He stood there and begged me to let him off. My punishment, he told me, not his own actions, would hurt his family. I flipped. He was the one who was hurting his family, not me. I stood up and slammed my fist so hard on my desk that the glass cover shattered with a great crash. My staff came running and rescued the sergeant, scarcely believing that their usually calm and cool commander had totally lost it. Frankly, it felt good, and I wasn't sorry to let them realize it could happen again.

I have occasionally exploded again, but I've never broken another glass desk top. I've learned how to display extreme, out-of-my-comfort-zone displeasure without destroying government property.

In the "heat of battle"—whether military or corporate—kindness, like calmness, reassures followers and holds their confidence. Kindness connects you with other human beings in a bond of mutual respect. If you care for your followers and show them kindness, they will reciprocate and care for you. They will not let you down or let you fail. They will accomplish whatever you have put in front of them.

11. HAVE A VISION. BE DEMANDING.

Followers need to know where their leaders are taking them and for what purpose. Mission, goals, strategy, and vision are conventional terms to indicate what organizations set out to accomplish. These are excellent and useful words, but I have come to prefer another and I believe better term—*purpose*. Think how often you see it—"sense of purpose" . . . "What's the purpose?" . . . "It serves a purpose."

Purpose is the destination of a vision. It energizes that vision, gives it force and drive. It should be positive and powerful, and serve the better angels of an organization.

Leaders must embed their own sense of purpose into the heart and soul of every follower. The purpose starts from the leader at the top, and through infectious, dynamic, passionate leadership, it is driven down throughout the organization. Every follower has his own organizational purpose that connects with the leader's overall purpose.

I once watched a TV documentary about the Empire State Building. For most of the hour, the documentary toured the wonders of the building—its history and structure: how many elevators it had, how many people worked or visited there, how many corporate offices it had, and how it was built. But at the end the story took a sharp turn. The last scene showed a cavernous room in a subbasement filled with hundreds of black trash bags, the building's daily detritus. Standing in front of the bags were five guys in work clothes. Their job, their mission, their goal was to toss these bags into waiting trash trucks.

The camera focused on one of the men. The narrator asked, "What's your job?" The answer to anyone watching was painfully obvious. But the guy smiled and said to the camera, "Our job is to make sure that tomorrow morning when people from all over the world come to this wonderful building, it shines, it is clean, and it looks great." His job was to drag bags, but he knew his purpose. He didn't feel he was just a trash hauler. His work was vital, and his purpose blended into the purpose of the building's most senior management eighty floors above. Their purpose was to make sure that this masterpiece of a building always welcomed and awed visitors, as it had done on opening day, May 1, 1931. The building management can only achieve their purpose if everyone on the team believes in it as strongly as the smiling guy in the subbasement.

Good leaders set vision, missions, and goals. Great leaders inspire every follower at every level to internalize their purpose, and to understand that their purpose goes far beyond the mere details of their job. When everyone is united in purpose, a positive purpose that serves not only the organization but also, hopefully, the world beyond it, you have a winning team.

Not long ago I spoke at a conference for the leaders of a credit rating company. Their whole focus seemed to be on reducing losses, eliminating high-risk applicants, purging bad debt, and speeding up the process. These goals are all essential to the success of the company, I told them, but they are all negative and hardly inspiring. Isn't your real purpose to find the right people to give credit to? Isn't your purpose to help people buy homes, educate their children, plan for their future? Isn't that what this conference should be all about?

Google's corporate mission statement is identical with its purpose: "to organize the world's information and make it universally accessible and useful." The founders set out to serve society, and created a remarkably successful company.

To achieve his purpose, a successful leader must set demanding standards and make sure they are met. Followers want to be "in a good outfit," as we say in the Army. I never saw a good unit that wasn't always stretching to meet a higher standard. The stretching was often accompanied by complaints about the effort required. But when the new standard was met, the followers celebrated with high-fives, pride, and playful gloating.

Standards must be achievable (though achieving them will always require extra effort), and the leaders must provide the means to get there. The focus should always be on getting better and better. We must always reach for the better way.

12. DON'T TAKE COUNSEL OF YOUR FEARS OR NAYSAYERS.

This one has a long history. You can trace it back to Marcus Aurelius, Andrew Jackson, Theodore Roosevelt, Winston Churchill, and hundreds of others. Perhaps the best known comes from Franklin D. Roosevelt's first inaugural address: "The only thing we have to fear is fear itself."

Fear is a normal human emotion. It is not in itself a killer. We can learn to be aware when fear grips us, and can train to operate through and in spite of our fear. If, on the other hand, we don't understand that fear is normal and has to be controlled and overcome, it will paralyze us and stop us in our tracks. We will no longer think clearly or analyze rationally. We prepare for it and control it; we never let it control us. If it does, we cannot lead.

I will never forget my fear the first time I came under fire. In 1963 I was the advisor to a Vietnamese infantry battalion. We were walking in column down a forested trail when we were hit by small arms fire from an enemy ambush. We returned fire and the Viet Cong enemy quickly melted back into the forest. It was over in a minute; but one soldier was killed. We wrapped him in a poncho and carried him until we found a place to bring in a helicopter. That night, as I tried to sleep on the forest floor, I was filled with the realization that the next morning we would probably be ambushed again. And we were. My body was filled with the gut feeling that I could be the next one killed. I was taller than the Vietnamese, and as the American advisor, I was a more valuable target. I stuck out.

That morning, and every morning, I had to use my training and self-discipline to control my fear and move on—just like all the Vietnamese, just like every soldier since ancient times. Moreover, as a leader, I could show no fear. I could not let fear control me.

Naysayers are everywhere. They feel it's the safest position to be in. It's the easiest armor to wear . . . And they may be right in their negativity; reality may be on their side. But chances are very good that it's not. You can only use their naysaying as one line in the spectrum of inputs to your decision. Listen to everyone you need to, and then go with your fearless instinct.

Each of us must work to become a hardheaded realist, or else we risk wasting our time and energy pursuing impossible dreams. Yet constant naysayers pursue no less impossible dreams. Their fear and cynicism move nothing forward. They kill progress. How many cynics built empires, great cities, or powerful corporations?

13. PERPETUAL OPTIMISM IS A FORCE MULTIPLIER.

In the military we are always looking for ways to leverage up our forces. Having greater communications and command and control over your forces than your enemy has over his is a force multiplier. Having greater logistics capability than the enemy is a force multiplier. Having better-trained commanders is a force multiplier.

Perpetual optimism, believing in yourself, believing in your purpose, believing you will prevail, and demonstrating passion and confidence is a force multiplier. If you believe and have prepared your followers, the followers will believe.

Late one winter's night in Korea after a very tough week of field training, my battalion of five hundred soldiers was waiting for trucks to take us back to our barracks at Camp Casey, twenty miles away. Word came down that we had a fuel shortage and no trucks were coming. We had to march back that night. The troops were exhausted, but we saddled up and started marching cross-country, with some grumbling in the ranks about higher headquarters.

After we launched, my operations officer, Captain Skip Mohr, reminded me that we had an outstanding requirement to make a forced twelve-mile timed march to qualify our troops to participate in the Expert Infantryman's Badge competition. He had plotted it out on the map; we would be twelve miles out in about half an hour. "Let's pick up the pace and go for it," he told me.

"Will that be pushing them too hard?" I wondered out loud.

"You know these kids," he answered. "They are tough as hell and will do anything we ask of them. They can do it."

I knew he was right.

We paused just before the twelve-mile point, took a ten-minute break, loosened our winter clothing, and then went for it, over some terrible hills. It was tough going. I wasn't sure I could keep up with these younger soldiers. But I pushed it, and so did they, magnificently. At the last mile, we could look down at the lights of Camp Casey. We fell into step and marched into camp in the middle of the night singing out a cadence and waking up everybody in the camp.

It was a great night. We had demanded a lot from our soldiers. But we had prepared them, we believed in them, they believed in us, and we had the confidence and optimism that they would succeed.

PART II

Know Yourself, Be Yourself

Chapter Two
Always Do Your Best, Someone Is Watching

Back when I was a teenager in the Bronx, summer was a time for both fun and work. Starting at about age fourteen, I worked summers and Christmas holidays at a toy and baby furniture store in the Bronx. The owner, Jay Sickser, a Russian Jewish immigrant, hired me off the street as I walked past his store. "You want to make a few bucks unloading a truck in back?" he asked me. I said yes. The job took a couple of hours, and he paid me fifty cents an hour. "You're a good worker," he told me when I'd finished. "Come back tomorrow."

That was the beginning of a close friendship with Jay and his family that continued through college and for the next fifty years, long after Jay had died. I worked part-time at the store a few hours a day during the summer and long hours during the Christmas season. I worked hard, a habit I got from my Jamaican immigrant parents. Every morning they left early for the garment district in Manhattan, and they came home late at night. All my relatives were hard workers. They came out of that common immigrant experience of arriving with nothing, expecting that the new life ahead of them would not be easy. Jamaicans had a joke: "That lazy brute, him only have two jobs."

After I'd worked at Sickser's for a couple of years, Jay grew concerned that I was getting too close to the store and the family. One day he took me aside. "Collie," he told me with a serious look, "I want you should get an education and do well. You're too good to just be a schlepper. The store will go to the family. You don't have a future here." I never thought I did, but I always treasured him for caring enough about me to say so.

When I was eighteen I became eligible to get a union card, which meant I could get a full-time summer job with better pay (I continued to work at Sickser's during the Christmas season). I joined the International Brotherhood of Teamsters' Local 812, the Soft Drink Workers Union. Every morning I went downtown to the union hall to stand in line to get a day's work as a helper on a soft-drink truck. It was hard work, and I became an expert at tossing wooden twenty-four-bottle Coca-Cola cases by grabbing a corner bottle without breaking it.

After a few weeks, the foreman noticed my work and asked if I'd like to try driving a Coke truck. Since I was a teamster, I had a chauffer's license and was authorized to drive a truck. The problem was that I had never driven a truck in my life. But, hey, why not? It paid better.

The next morning, I got behind the wheel of an ancient, stick shift, circa 1940 truck with a supervisor riding shotgun. We carried three hundred cases, half on open racks on one side of the truck and half on the other. I asked the supervisor where we were going. "Wall Street," he said, and my heart skipped a beat as I imagined navigating the narrow streets and alleys of the oldest, most claustrophobic, and most mazelike part of New York City. I took off with all the energy and blind optimism of youth and managed to get through the day and somehow safely delivered the three hundred cases . . . in spite of my often overenthusiastic driving. My supervisor was white-knuckled with worry that I would deliver 150 cases onto the street as the old truck leaned precariously at corners I was taking much too fast. Though I delivered every case, my driving skills did not impress the supervisor, and my truck-driving career was over (they still kept me on as a helper). Nevertheless, I proudly took home a $20 salary that day to show my father.

The next summer, I wanted something better than standing in a crowd every morning hoping for a day's work. My opportunity came when the hiring boss announced one morning that the Pepsi plant in Long Island City was looking for porters to clean the floors, full-time for the summer. I raised my hand. I was the only one who did.

The porters at the Pepsi plant were all black. The workers on the bottling machines were all white. I didn't care. I just wanted work for the summer, and I worked hard, mopping up syrup and soda that had spilled from overturned pallets.

At summer's end, the boss told me he was pleased with my work and asked if I wanted to come back. "Yes," I answered, "but not as a porter." He agreed, and next summer I worked on the bottling machine and as a pallet stacker, a more prestigious and higher-paying job. It wasn't exactly the Selma March, but I integrated a bottling machine crew.

Very often my best didn't turn out that well. I was neither an athlete nor a standout student. I played baseball, football, stickball, and all the other Bronx sports, and I did my best, but I wasn't good at any. In school

I was hardworking and dedicated, but never produced superior grades or matched the academic successes of my many high-achieving cousins. Yet my parents didn't pester me or put too much pressure on me. Their attitude was "Do your best—we'll accept your best, but nothing less."

These experiences established a pattern for all the years and careers that came afterward. Always do your best, no matter how difficult the job, or how much you dislike it, your bosses, the work environment, or your fellow workers. As the old expression goes, if you take the king's coin, you give the king his due.

I remember an old story told by the comedian Brother Dave Gardner about two ditch diggers. One guy just loves digging. He digs all day long and says nothing much. The other guy digs a little, leans on his shovel a lot, and mouths off constantly, "One of these days, I'm gonna own this company."

Time passes and guy number one gets a front-end trench machine and just digs away, hundreds of feet a day, always loving it. The other guy does the minimum, but never stops mouthing off, "One of these days, I'm gonna own this company." No, guy number one doesn't end up owning the company, but he does become a foreman working out of an air-conditioned van. He often waves to his old friend leaning on his shovel still insisting, "One of these days, I'm gonna own this company." Ain't gonna happen.

In my military career I often got jobs I wasn't crazy about, or I was put in situations that stretched me beyond my rank and experience. Whether the going was rough or smooth, I always tried to do my best and to be loyal to my superior and the mission given to me.

On my second tour in Vietnam, I was assigned as an infantry battalion executive officer, second in command, in the 23rd Infantry Division (Americal). I was very pleased with the assignment. As it happened, I had just graduated with honors from the Command and General Staff College at Fort Leavenworth, Kansas. Shortly after I arrived in Vietnam, a photo of the top five graduates appeared in *Army Times*. The division commanding general saw it, and I was pulled up to the division staff to serve as the operations officer, responsible for coordinating the combat operations of a twenty-thousand-man division. I was only a major and it was a lieutenant colonel's position. I would have preferred to stay with my battalion, but wasn't given that choice. It turned out to be very demanding and a stretch for me, but it marked a turning point in my career. Someone was watching.

Years later, as a brigadier general in an infantry division, I thought I was doing my best to train soldiers and serve my commander. He disagreed and rated me below standards. The report is still in my file. It could have ended my career, but more senior leaders saw other qualities and capabilities in me and moved me up into more challenging positions, where I did well.

Doing your best for your boss doesn't mean you will always like or approve of what he wants you to do; there will be times when you will have very different priorities from his. In the military, your superiors may have very different ideas than you do about what should be your most important mission. In some of my units my superiors put an intense focus on reenlistment rates, AWOL rate, and saving bonds participation. Most of us down below would have preferred to keep our primary focus on training. Sure, those management priorities were important in principle, but they often seemed in practice to be distractions from our real work. I never tried to fight my superiors' priorities. Instead I worked hard to accomplish the tasks they set as quickly and decisively as I could. The sooner I could satisfy my superiors, the sooner they would stop bugging me about them, and the quicker I could move on to my own priorities. Always give the king his due first.

By the end of my career in government, I had been appointed to the nation's most senior national security jobs, National Security Advisor, Chairman of the Joint

Chiefs of Staff, and Secretary of State. I went about each job with the same attitude I'd had at Sickser's.

During my tenure as Secretary of State, I worked hard on President Bush's agenda, and we accomplished a great deal that has not received the credit it deserved. We forged good relationships with China, India, and the Russian Federation, all major powers and all potential political adversaries. We did historic work on disease prevention in the Third World, including HIV/AIDS, and we significantly increased aid to developing countries. In the aftermath of 9/11, we made the nation more secure. We got rid of the horrific Hussein and Taliban regimes in Iraq and Afghanistan. But the residual problems in those countries exposed deep fissures within the national security team. By the beginning of 2004, our fourth year, the Bush national security team had in my view become dysfunctional, which has been well documented. Since it was obvious that my thinking and advice were increasingly out of sync with the others on the team, the best course for me was to leave. At that time I strongly believed that for his second term the President should choose an entirely new national security team, and I gave that advice to President Bush, but he chose not to take it. I left the State Department in January 2005. President Bush and I parted on good terms.

In the years that have followed my government service, I have traveled around the country and shared my life's experience with many people in many different forums. At these events, I always emphasize, especially to youngsters, that 99 percent of work can be seen as noble. There are few truly degrading jobs. Every job is a learning experience, and we can develop and grow in every one.

If you take the pay, earn it. Always do your very best. Even when no one else is looking, you always are. Don't disappoint yourself.

Chapter Three
The Street Sweeper

I have always tried to keep my life in perspective with my ego under control. That effort has been helped enormously by a wife and three kids who have never taken me too seriously and who have always held above me an imaginary oxygen mask ready to drop down whenever I needed a whiff of reality. The first time I came home looking sharp in the new battle dress camouflage fatigues the Army adopted in the 1980s, my daughter Annemarie, then about twelve, merely looked up from watching television and announced, "Mom, the GI Joe doll is home."

Over time, others have helped me keep my ego down. After I retired, I was invited to give a speech to a large luncheon event in Boston. There were about two thousand guests and you needed two tickets, one to get into the room, and the second for the waitress to verify that you had paid for lunch. I was escorted to the round head table by the event's chairman. As the waitress placed salads before each guest, she asked for meal tickets. She passed by me without giving me a salad. When it was time for the next course she passed me by again. That was when the chairman realized what was going wrong. Mortified, he said to the waitress, "Young lady, this is General Colin Powell, former Chairman of the Joint Chiefs of Staff, our honored guest and keynote speaker." Her simple, no-nonsense response was "He ain't got no ticket, huh?" The chairman produced a ticket for me. I was getting hungry.

I love when people do their job. Doing your job well, with someone watching, without inflating your self-importance or showing off, is not easy.

Some years ago, there was a human interest segment about a street sweeper on the evening news. I think he worked in Philadelphia. He was a black gentleman and swept streets the old-fashioned way, with one of those wide, stiff bristle brooms and a wheeled garbage can. He had a wife and several children and lived in a modest home. It was a loving family, and he had high ambitions for his children. He enjoyed his job very much and felt he was providing a worthwhile service to his community. He had only one professional ambition in life and that was to get promoted to drive one of those mechanized street sweepers with big round brushes.

He finally achieved his ambition and was promoted to driving a street sweeping machine. His wife and children were proud of him. The television piece closed with him driving down the street; a huge smile was on his face. He knew who he was and what he was.

I run that video piece through my mind every few months as a reality check. Here is a man happy in his work, providing an essential service for his community, providing for his family, who love and respect him. Have I been more successful in what is truly important in life than he has been? No, we have both been fortunate. He has touched all the important bases in the game of life. When we are ultimately judged, despite my titles and medals, he may have a few points on me, and on a lot of others I know.

Chapter Four
Busy Bastards

The 23rd Infantry Division (Americal), where I served in Vietnam for a short time as operations officer, was commanded by a wonderful soldier, Major General Charles M. Gettys. I learned a great deal from General Gettys. He was a calm, confident commander, not given to outbursts or showing off his rank. He placed great confidence in his staff, but there was no question who was in charge.

He and I were casually chatting one day when the name of another general came up. He was a highly regarded officer, but Gettys had reservations about him. "Colin, he's a good guy," he told me, "but he is one of those 'busy bastards.' He always has to be doing things and coming up with new ideas and working absurd hours."

Gettys's wisdom has stayed with me, and I have tried to learn from it. He pointed out back then (maybe intentionally) a road I was inclined to travel. I've always done my best to come up with new ideas, and I certainly worked hard in all my jobs. But I have tried not to be a busy bastard. As President Reagan used to frequently observe, "They say hard work never killed anyone, but why take a chance?"

I've seen many busy bastards over the years . . . I shouldn't call them bastards, but Gettys's words have burned into my brain. Most of them are good people, not bastards. They just can't ever let it go.

A busy bastard never leaves the office until late at night. He has to go in on weekends. He shows up in the morning at hours suitable only for TV traffic announcers, failing to recognize that a couple dozen staff people have to show up at the same time to make sure he gets the support he can't do without and to prove they're as committed to the job as he is.

In every senior job I've had I've tried to create an environment of professionalism and the very highest standards. When it was necessary to get a job done, I expected my subordinates to work around the clock. When that was not necessary, I wanted them to work normal hours, go home at a decent time, play with the kids, enjoy family and friends, read a novel, clear their heads, daydream, and refresh themselves. I wanted them to have a life outside the office. I am paying them for the quality of their work, not for the hours they work. That kind of environment has always produced the best results for me.

I tried to practice what I preached. I enjoy fixing things, especially old cars, and especially old Volvos. The Chairman of the Joint Chiefs of Staff lives across the river from Washington in a mansion in Fort Myer on a hill overlooking the city. A hundred feet behind the mansion were three garages. When I was Chairman, the garages were always filled with dead circa-1960 Volvos waiting to be fixed or stripped for parts. People who really needed to see me on weekends knew where to find me . . . under a Volvo. If they wanted to visit or chat, I didn't mind, as long as I could continue working. I enjoyed analyzing a dead engine to discover why it wouldn't start, reducing the possibilities for the failure down to one, fixing it, and then rejoicing when the engine fired up. My office problems seldom lent themselves to such straightforward, linear analysis. Once a car was running, I had no further interest in it. I would buy a ninety-nine-dollar Earl Scheib paint job and sell it as fast as possible. I was under a Volvo one Sunday in 1989 during our invasion of Panama when the Operations

Center called to tell me we had picked up the dictator Manuel Noriega.

While I was making the transition to Secretary of State, I interviewed a number of candidates for senior positions. Toward the end of one of these interviews, an extremely able and gifted Foreign Service officer asked if I would mind if he went out to jog in the afternoons.

"You can go home and jog as far as I'm concerned," I told him. "I trust you to know how to get your work done without me maintaining a sign-out sheet on you."

The very fact that a senior officer would ask such a question pointed out how necessary it was to demonstrate to my staff that I wasn't a busy bastard.

My mentor in this style of operating was Frank Carlucci. When the Reagan administration took office in 1981, Frank was appointed Deputy Secretary of Defense; and I became his military assistant. Because Frank always tried to leave the office at a reasonable hour and avoided the place like a plague on weekends, I worked reasonable hours and so did everyone else on his staff. We ran a very efficient office.

In the spring of 1981, I persuaded Frank to release me for a field assignment. The officer who replaced me, a compulsive worker, stayed late every night. Even though Frank only rarely came in on weekends, and never for more than a couple of hours, his new military assistant felt he had to be there. Sure enough, all those extra hours generated more work for the entire staff. The workload expanded to fill the time. Most of it was make-work, anything but necessary or important. Frank found himself with additional paper he didn't ask for, need, or expect. He had to start working longer hours!

In late 1986, in the aftermath of the Iran-Contra scandal, Frank became President Reagan's National Security Advisor and I became Frank's deputy. Our task was to reorganize the national security system and fix the deficiencies that had caused the scandal. Even during this stressful, demanding time, with a presidency at risk, Frank maintained his long-standing work habits. One of my responsibilities as his deputy was to keep an eye on him to make sure he didn't have to work late. I didn't have to worry. Left to his own devices, with no crisis pending, he would leave at 3 p.m., play tennis, and go home. He worked hard, was incredibly well organized, and got the work done. The staff followed Frank's lead.

By the time I had reached my most senior positions, I never went to the office on weekends unless a war had just started or some other crisis demanded my presence. On Fridays, I left the office with tons of work; I was far more efficient in the quiet privacy of my home.

I expected my staff to do likewise. If you have a reason to go in, then go in, but never think that going in just for the sake of going in impresses me.

President Reagan was a joy in this regard. He didn't need encouragement to keep reasonable work hours. When Frank Carlucci became Secretary of Defense, I took over as National Security Advisor. As I'd done earlier with Frank, one of my jobs was to watch the President's schedule to make sure we didn't keep him late. Toward the end of the day, we gave him a homework package. He was normally upstairs in the residence with Mrs. Reagan by six o'clock. Friday afternoons were even better. Right after lunch, he usually got an end-of-the-week briefing from Secretary of State George Shultz. Reagan would listen patiently but with limited attention. Around 2:15, when he heard the drone of Marine One descending onto the South Lawn, he'd perk up. It was time to leave for Camp David! He'd arrive there by 3 p.m., and short of an emergency, stay until Sunday evening. Seldom were guests invited to Camp David. The President relaxed, read staff papers and books, and spent time with Mrs. Reagan. This was their time. And, hallelujah, it was our time to get caught up, spend time with our families, and rest up and get ready for the demanding week ahead. The nation was safe without the President whizzing all over the place on weekends. Our only concern was the books he was reading. Despite our best efforts, old friends would now and again slip seriously odd books into his briefcase, generating often unanswerable questions Monday morning. One Monday, the President came in brimming over with curiosity about how trees create pollution.

Reagan loved relaxing at his ranch in the Santa Ynez Mountains just outside Santa Barbara, California. We loved it even more. We were condemned to camp out in fancy cabana suites on the beach at the beautiful Santa Barbara Biltmore hotel. Twice a day the senior staff assembled to see what we needed to tell him. We'd telephone up to the ranch and brief him, and we'd send up intelligence, situation reports, and papers for him to work on. If no crises were looming, we could quietly take care of business and prepare for the challenges ahead or split for the pool or the beach, making sure we could monitor everything in case of an emergency. It was rare for anyone to have to brief him at the ranch. I went up just once, to brief him on a treaty we had just concluded with the Russians to reduce our nuclear weapons inventories.

I worked hard all my life and always expected those who worked for me to do likewise. But I tried not to generate make-work. I learned early that a complete life includes more than work. We need family, rest, outside interests, and time to pursue them. I always keep in mind a lesson taught to all young infantry lieutenants: "Don't run if you can walk; don't stand up if you can sit down; don't sit down if you can lie down; and don't stay awake if you can go to sleep."

Chapter Five
Kindness Works

Many years ago I was the warden—the senior lay person—of a small suburban Episcopal church in northern Virginia. During that time our bishop assigned to our parish an elderly priest to serve as an assistant pastor. The priest was in some kind of personal distress and needed a parish home. I never knew the nature of his problem. Whatever it was, we were pleased to take him in. We welcomed him into the church family, treated him as one of us, and ministered to him, just as we ministered to each other. Nobody asked about his problem or pried into his life.

He was with us for a year. On his last Sunday he was assigned to give the sermon. I listened to it in my usual proper Episcopalian position, right rear of the church. I'm sure it was a good sermon, but one sentence hit me with special force and has remained with me for four decades. At the end of the sermon, the priest looked over the congregation and with a smile on his face quietly concluded: "Always show more kindness than seems necessary, because the person receiving it needs it more than you will ever know."

He was talking about himself, of course. The lesson was clear: Don't just show kindness in passing or to be courteous. Show it in depth, show it with passion, and expect nothing in return. Kindness is not just about being nice; it's about recognizing another human being who deserves care and respect.

Much later, when I was Secretary of State, I slipped away one day from my beautiful office suite and vigilant security agents and snuck down to the garage. The garage is run by contract employees, most of them immigrants and minorities making only a few dollars above minimum wage.

The garage is too small for all the employees' cars. The challenge every morning is to pack them all in. The attendants' system is to stack cars one behind the other, so densely packed that there's no room to maneuver. Since number three can't get out until number one and two have left, the evening rush hour is chaos if the lead cars don't exit the garage on time. Inevitably a lot of impatient people have to stand around waiting their turn.

The attendants had never seen a Secretary wandering around the garage before; they thought I was lost. (That may have been true by then, but I'd never admit it.) They asked if I needed help getting back "home."

"No," I answered. "I just want to look around and chat with you." They were surprised, but pleased. I asked about the job, where they were from, were there problems with carbon monoxide, and similar small talk. They assured me everything was fine, and we all relaxed and chatted away.

After a while I asked a question that had puzzled me: "When the cars come in every morning, how do you decide who ends up first to get out, and who ends up second and third?"

They gave each other knowing looks and little smiles. "Mr. Secretary," one of them said, "it kinda goes like this. When you drive in, if you lower the window, look out, smile, and you know our name, or you say 'Good morning, how are you?' or something like that, you're number one to get out. But if you just look straight ahead and don't show you even see us or that we are doing something for you, well, you are likely to be one of the last to get out."

I thanked them, smiled, and made my way back to where I had abandoned my now distraught bodyguard.

At my next staff meeting, I shared this story with my senior leaders. "You can never err by treating everyone in the building with respect, thoughtfulness, and a kind word," I told them. "Every one of our employees is an essential employee. Every one of them wants to be viewed that way. And if you treat them that way, they will view you that way. They will not let you down or let you fail. They will accomplish whatever you have put in front of them."

It ain't brain surgery. Every person in an organization has value and wants that value to be recognized. Every human being needs appreciation and reinforcement. The person who came to clean my office each night was no less a person than the President, a general, or a cabinet member. They deserved and got from me a thank-you, a kind word, an inquiry that let him or her know their value. I wanted them to know they weren't just janitors. I couldn't do my job without them, and the department relied on them. There are no trivial jobs in any successful organization. But there are all too many trivial leaders who don't understand this oh so simple and easy to apply principle.

Taking care of employees is perhaps the best form of kindness. When young soldiers go to basic training they meet a drill sergeant, who seems to be their worst nightmare. He shouts at them relentlessly, he intimidates them, he makes them miserable. They are terrified. But all that changes. Their fear and initial hatred turn into something else by the end of basic training. The sergeant has been with them every step of the way: teaching, cajoling, enforcing, bringing out of them strength and confidence they didn't know they had. At the end, all they want is for their performance to please him. When they graduate, they leave with an emotional bond and a remembrance they will never forget. Ask any veteran the name of his drill sergeant and he will know it. My ROTC summer camp drill sergeant almost fifty-five years ago was Staff Sergeant (SSG) Artis Westberry.

Being kind doesn't mean being soft or a wuss. Kindness is not a sign of weakness. It is a sign of confidence. If you have developed a reputation for kindness and consideration, then even the most unpleasant decisions will go down easier because everyone will understand why you are doing what you are doing. They will realize that your decision must be necessary, and is not arbitrary or without empathy.

As the old saying puts it, "To the world, you may be one person, but to one person you may be the world."

Chapter Six
I'm All Caught Up

One of my early mentors, Captain Tom Miller, a wonderful man, commanded Company B, 2nd Armored Rifle Battalion, 48th Infantry in Germany in the late 1950s. I was one of his lieutenants. It was my first assignment. Tom was one of several World War II and Korean War veterans commanding companies in those days, mostly reservists or sergeants who had been promoted during the wars. None of them was destined to be a general, but, boy, they knew a lot about soldiering.

We didn't call it mentoring back then. It was just what senior officers were supposed to do—train and guide young lieutenants just starting out and try to keep them out of trouble until they were weaned. We learned a lot during the day, but the learning that took place at the officers' club bar at night was a lot more important and a lot more fun.

Late one night Captain Miller and several lieutenants were sitting at the bar drinking beers. We'd all had more than one, but Tom was way ahead of us as usual. He looked over at us and said, "Now, listen you guys, I wanna tell you about leadership. You all think you are pretty sharp. And at the end of the day you leave the company thinking you've got everything in great shape. All the rifles are accounted for, no troops are AWOL, everyone has made bed check, and you've had a good day of training. You think everything is squared away. You're patting yourselves on the back. Then, in the middle of the night, when no one is looking, things get bad screwed up. The next morning you discover a fight had broken out, four windows are broken, two guys are in the hospital, one is missing, a jeep is gone, and the MPs are there waiting for you. You know what? You just suck it up and get started again. It's a new day in which to excel."

I had many mornings like that over the next fifty years. We all do. Problems come with just being alive, and even more come with responsibility. When they come, you just suck it up and get started again. You are never caught up. I've lived by the proposition that solving problems is what leaders do. The day you are not solving problems or are not up to your butt in problems is probably a day you are no longer leading. If your desk is clean and no one is bringing you problems, you should be very worried. It means that people don't think you can solve them or don't want to hear about them. Or, far worse, it means they don't think you care. Either way it means your followers have lost confidence in you and you are no longer their leader, no matter what your rank or the title on your door.

So go walk around and look for a problem; you will find some.

Don't stop there. Try to instill a problem-solving attitude in your subordinates and staff.

In 1973, I was a battalion commander in Korea. One day I lit into all my commanders and senior sergeants about problems that kept popping up with the troops. I didn't think my leaders were watching and listening closely enough to the troops, and I let them know I wasn't happy. Later that afternoon I was taking my customary walk through the battalion area. As I came around the back of a Quonset hut, I heard SSG Walker, one of my best noncommissioned officers (NCOs), talking to his platoon in formation. It went something like this: "Now listen up! I got chewed out this morning by the CO about your problems. That ain't gonna happen again. Now, if any of you clowns got a problem I want you to fall out and meet me in my hootch to tell me what it's all about and I'm gonna solve it right now. Any questions?" I shook my head, laughing. SSG Walker's troops seldom had problems he didn't know about.

I'm a restless guy. I like to move. I don't like spending long periods of time at my desk. In all my assignments, from lieutenant to Secretary of State, I always spent time going on walkabout, as our Australian friends call it. Sometimes I would wander around with no particular route in mind, and would show up in unexpected places—the State Department boiler room, for example, or the Pentagon Police Station. In my commands, I sometimes wandered where the spirit moved me and sometimes I followed precise paths through troop areas at predictable times. Junior officers, NCOs, and troops knew when and where they could ambush me with their problems. I found out things that would never or not easily flow through the staff or up the chain of command.

I followed up on every problem I got, but did it in a way that didn't undercut the chain of command. I tried to make sure my subordinates knew not to be threatened by my roaming around, and I gave them first shot at solving the problem . . . unless they were the problem.

Problems have to be solved, not managed. You can't get away with burying them, minimizing them, reorganizing around them, softening them, or assigning blame somewhere outside your responsibility. You have to make real and effective changes. You can't fool a GI, you can't fool a floor worker, and you can't fool a store cashier. They know when something is wrong, and they know it first. They know when someone is not a good follower, not getting the job done. They are waiting for you to find out and do something about it. If you don't, they will start slacking off. If you don't see it, or having seen it, don't care enough to do something, why should they care about you? Good followers who know you care not only do a good job, they take care of you.

There is a very old story from the days before Amtrak when we had passenger railroads all over the country. One day the president of the New York Central Railroad got an outraged letter from an irate passenger who'd taken a sleeper from New York City to Buffalo. The bed was full of bedbugs. Within a week, the passenger got a profusely apologetic letter from the president. "We greatly value your patronage," it said. "We promise to have the problem fixed." The passenger was momentarily satisfied . . . until he read the handwritten note from the president to his secretary that had slipped out of the envelope. It said: "Send this jerk the 'bedbug letter.' "

I have thrown a lot of unsigned letters into my outbox over the years. "Solve the problem," I've told my staffs again and again. "I don't do bedbug letters."

Chapter Seven
Where on the Battlefield?

Shortly after I became Secretary of State, I received an insightful—and surprising—letter from Ambassador George Kennan, the Grand Old Man of American Diplomacy. I had never met Ambassador Kennan, but I knew him as the most highly regarded, influential, and prophetic American diplomat of the last century. A letter from Kennan was like a report from the burning bush by the Moses of diplomats. When I opened it I expected wise commentary on the great geostrategic issues of the day. Instead, he gave me three pages of heartfelt advice about my new job.

Though then ninety-seven (he died, aged 101, in 2005), he could still produce clear, succinct, powerfully argued prose. As if I needed it, he began by establishing his credentials—oldest living member of the original Foreign Service of 1925–75; seventy-five years of foreign affairs experience as a diplomat and historian; protégé of George Marshall; one of the chief architects of the plan that bears Marshall's name; and author of the famous "Long Telegram" from Moscow, which laid the foundation of the containment policy that shaped America's strategy toward the Soviet Union until it collapsed. Kennan was a man of strong opinions and a speaker of hard, unpalatable truths, a lone voice driven more than once into the wilderness. He was always revered, but not always listened to.

After his personal history came the heart of the letter, which started with a reminder of the Founding Fathers' intention in the years after our nation's birth regarding the two principal duties of the Secretary of State. The first was to function as the President's most intimate and authoritative advisor on all aspects of American foreign policy. The second was to exercise administrative control over the State Department and the Foreign Service. He then cut to the chase: you can't properly perform either of these duties if you are constantly running around the world in your airplane. Recent Secretaries of State, in his view, had been spending too much time flying to other countries for face-to-face meetings with foreign leaders and dignitaries. The role of the Secretary of State is principal foreign policy advisor to the President, not highest-ranking roving ambassador. Surely modern communications made it possible to conduct diplomacy without flying off to meetings all over the world. He had no quarrel with brief travel away from Washington when official duty required it. But absences should be held to a minimum and avoided when suitable alternatives were available.

The problem of Secretaries traveling too much, he continued, was not limited to questions about his presence or absence in Washington. Ambassadors are the President's representatives to the other nations of the world—the official, institutional, government-to-government links between countries. Because he is there every day, the ambassador's position should be enhanced as the main channel of diplomatic activity. The too-frequent arrival of the Secretary and assorted special envoys tends to undercut that role. Why spend time with the ambassador when you can persuade the Secretary to drop by?

Well, the Kennan letter pretty much matched the way I wanted to approach the job, and I embraced its recommendations. In my four years as Secretary I traveled a great deal, but not as much as some of my predecessors and nowhere near as much as my successors. Condoleezza Rice and Hillary Clinton set world records.

For some unknown reason, the media, led by the *New York Times*, started clocking my frequent flyer miles. I didn't travel enough, they claimed. I should be making more waves out there in the world rather than spending so much time in Washington or at the UN headquarters in New York.

None of them answered the obvious question: Is this trip really necessary? What national purpose is served by having me out there? And none asked me if I might have good reasons for remaining in Washington.

Truth is, in my first year I traveled to thirty-seven countries and logged 149,000 miles . . . not exactly hiding in a bunker.

For years I've been a frequent traveler. Even today I'm on the road as much as 50 percent of the time. But I don't long to travel. Years ago travel lost any glamour it may have had. I travel for work, not for pleasure. Any trip I take has to be necessary. It has to have a purpose and a function. I am not by nature a good tourist; I've seen most of the world's sights that I've wanted to see. When I was Secretary, I met with leaders, visited schools, talked to kids, and was a spectator at cultural events, but I seldom lingered to tour and shop. I used telephones, the then-newfangled email system, and cell phones to stay in touch with my foreign counterparts around the world. I attended every NATO and European Union meeting, every official gathering of Asian leaders, every Organization of American States meeting, and made more trips to Africa than any of my predecessors.

In fact, during my four years as Secretary some of the biggest problems and decisions made back in D.C. occurred when I was twelve hours away in some hotel overseas. I was in Peru on 9/11. I was in Asia when important decisions were made concerning our detention and interrogation policies. I wish I had been in Washington at such times.

My way of managing my time and travel is not the only way. Other Secretaries may have the better argument. Today more frequent overseas travel may be a better and more appropriate use of a Secretary's time than watching over State Department business in Washington. The presence of the Secretary in other countries shows the flag in a very special way. This in itself can be as important as private meetings or attending conferences. The world has changed since the eighteenth century. Travel between countries now takes hours, not weeks or months. Face-to-face presence is easy. We all have to adapt to the age we live in. One could argue that Kennan was trying to bring back to life a vanished age.

There is no single best way to do the job. Every Secretary and, for that matter, every leader has to make a judgment about where to focus his efforts.

The right answer for a Secretary of State is, of course, to balance the requirements to participate in international forums, maintain bilateral comity with other nations, and be present to run a large department and serve the President. Deputies, assistants, staff, and communications help, but the leader can only be physically in one place at a time. And physical presence trumps electronic presence.

My own solution to the problem of finding the right balance has been shaped by my military training and experience. In the military the problem is posed this way: "Where should the commander be on the battlefield?" The answer: "Where he can exercise the greatest influence and be close to the point of decision"—the place where personal presence can make the difference between success and failure. A battalion commander leading a charge up a hill with seven hundred troops behind him may be a courageous and inspirational figure, but he is at that moment just another infantryman trying to stay alive. He can't see the whole battlefield; he is not in a position to move forces; he can't communicate with all his subordinates, arrange more support, or keep higher headquarters informed. The battalion commander who is firing a rifle and no longer commanding his battalion is, as we say, "decisively engaged." (A commander is decisively engaged when he is in a win-or-lose situation and has lost freedom of movement.)

Corporate leaders will of course have different answers to the "Where on the battlefield?" question than military leaders or Secretaries of State. But for each of them the answer has to be "at the point of decision." The point of decision can be many places. Because it is important for followers to see and hear from their leader, corporate executives should often visit the factory floor to see what is going on. But then get out of the way so workers, foremen, and line leaders can get on with their jobs. Get back upstairs and work to make sure the guys downstairs get what they need to do the job. That's what you're being paid for!

The point of decision might be a television show explaining to the world the revolutionary new product you are getting ready to unveil (see Steve Jobs) or why you overinvested in complex derivatives or subprime investments (too many to mention). Maybe you need to be up on Capitol Hill getting keelhauled by a first-term congressman.

There are lots of recent examples of executive failure to be at the point of decision. During the 2008 and 2009 economic recession, we saw CEOs at bridge tournaments or playing golf while all hell was breaking loose in their corporate headquarters. They were neither in a place to influence the action nor in a decisive position to win the battle.

I watched with profound disbelief as the top executives at Lehman Brothers again and again sent out a new and inexperienced chief financial officer to explain why their company was getting sucked into a black hole, while they sequestered themselves in their paneled offices.

The right answer to "Where on the battlefield?" is a function of a leader's experience, self-confidence, confidence in his subordinates, and the needs of his superiors. In my career, I constantly asked myself where my point of decision was—the best place to see what is really going on, to influence the outcome, and to retain freedom of movement. During Operation Desert Storm, I only occasionally visited General Schwarzkopf at his headquarters in Riyadh. My place was in the Pentagon making sure he and his half a million troops got what they needed, not the least of which was political and public relations support.

One week into the war, the public mood had become unsettled and the media was becoming critical. After the success of the first day and the excitement of watching cruise missiles strike with incredible accuracy, it looked from the outside as though the war was going nowhere. "Why isn't it over?" people were asking.

Secretary of Defense Dick Cheney and I realized we had to act to settle things down. The point of decision for us at that moment was not in our offices or in situation rooms monitoring the war, but down in the press room. We called a press conference where Dick gave an excellent summary of the strategic and political situation, and then I covered the military campaign. I summarized our actions during the previous week, concluding with a few sharp words detailing our strategy to kick the Iraqi army out of Kuwait: "First we are going to cut it off," I told the assembled reporters, "and then we are going to kill it." My line was picked up by all the newspapers and all the radio and TV news shows. It did the trick. It told the people out there what they needed to know. Confidence about our war aims returned. And Dick and I could leave the front lines and get back to our offices.

General George Marshall, Army Chief of Staff during World War II, wanted desperately to lead the D-Day invasion of Europe. Any general would want to lead the "Great Crusade." But that didn't happen. The assignment went to General Eisenhower, one of his protégés and junior to him. President Roosevelt, well aware of how badly Marshall wanted the mission, discussed it with him. At the end of the conversation, as Marshall was leaving, Roosevelt said gently,

"Well, I didn't feel that I could sleep at ease if you were out of Washington." Marshall, that great man, knew his place was not to wade into the surf out of a landing craft in the Philippines or command the assault on the Normandy beaches, but to ensure that MacArthur and Eisenhower could.

Chapter Eight
Spheres and Pyramids

Most organizations are like pyramids, with the leaders at the top and everybody else on descending layers down to the bottom, where the heaviest physical work normally takes place and where people start out. Now imagine each person in that organization to be a sphere. On the lowest level of the pyramid, each sphere is tiny, but it's capable of growing. Everything outside the pyramid is the environment in which the organization lives.

Over time people ascend within the pyramid from layer to ever higher and narrower layers. As they gain experience and show ability, their spheres grow bigger and bigger until they hit the inside walls of the pyramid. During this process, hopefully, they develop into leaders. Once that happens, the only way to keep growing and rising is to expand outside the pyramid. Rising leaders begin to learn about the world outside the narrow confines of the pyramid, the world in which the pyramid exists.

In most organizations leaders are chosen in one of two ways. They rise up from the bottom of the pyramid, or they are brought in from outside. In the military, we only grow our leaders inside the pyramid. If we need a battalion commander, he or she must have come up inside the organization. We don't lateral them in from IBM.

Because so much of their life is spent inside their own service, rising young officers may not have wide experience of the world outside their beloved and comfortable pyramid. Senior officers must learn about that world. They must also learn about the contributions of the other services to the nation's safety and security. They must gain experience in the operations of international alliances, such as NATO. They must understand and appreciate the political process, the role of Congress, the civilian departments of government, media relations, economics, and a host of other subjects outside the pyramid. Only when an officer has mastered these areas is he able to rise to the more senior layers, where increasingly he deals with and integrates that outside world. If a rising leader fails to understand

and engage with the outside environment his sphere will never expand beyond the walls of the pyramid and he will stop rising.

In the Army, this is how you rise up through and beyond the pyramid. Let's say you're a young infantry lieutenant. You start out as a tiny sphere deep in one of the corners at the bottom of the pyramid. Your job is to master both your corner of the pyramid and command of a platoon, taking care of forty soldiers. As a young lieutenant, you don't worry about geopolitical issues or how the economy is doing. Your life at that moment is dedicated to preparing yourself and those forty soldiers for battle.

Time passes, your sphere gets bigger, and you begin to rise in the organization. You become expert in your field, no longer an apprentice. You are of increased value to the pyramid.

More time passes . . . about fifteen years. By then you may have become a superb battalion commander. Your sphere has become so large that it starts to rub against the sides of the pyramid.

You are sent to more senior military schools that focus beyond the skills and knowledge a battalion commander must master. You learn how to lead larger, more complex organizations. You learn the importance of working with the other services. You may be sent to a civilian graduate school to get an MBA or other advanced degrees. You begin to work with higher level civilians, even politicians.

More years pass. As you rise higher, your personal sphere increasingly balloons outside the narrowing pyramid. You leave behind many of your peers, even though they are well qualified. Some have not grown to meet the increased expectations placed upon them; for others there's simply not enough room. Not everyone qualified will climb to the top of the pyramid.

You become a general. You no longer wear the insignia of your original branch, such as the crossed rifles of the infantry. When you are promoted to brigadier general they pin the star on you, give you a red flag with a single white star in the middle. You get a special General Officer's belt to wear in the field. You get a special edition General Officer's pistol. You are not in Kansas anymore. You are a vice president of the company.

More years pass. You may rise higher up the pyramid and gain more stars. You may never see another infantry unit. You might even rise to the very top, the Chairman of the Joint Chiefs of Staff, where you have responsibility to supervise other services and not just the Army. Your pyramid is no longer just the Army pyramid; it includes all the other military pyramids.

You are at the pinnacle of our military pyramid and can go no higher. At the top, most of your time is devoted to the external environment: relations with allies, working with international organizations, the White House. Your job is to seek opportunities, identify risks, obtain resources, and serve as the lead spokesman for the needs, aspirations, and purposes of all the services. You will find yourself connected to other pyramids—the intelligence pyramid, the economic pyramid, the budget pyramid.

If you get to the top, you have worked hard to know and improve yourself and to expand your vision beyond the constraints of the pyramid; opportunities have come your way; you have picked up a champion mentor; equals have left the pyramid; and you have been very lucky.

You are probably not entirely comfortable perched up there on the tip of the pyramid. There are myriad competing demands, pressures, and gut-wrenching decisions. Mistakes have large consequences. You are a highly visible target. It is easy to fall off. In spite of these pressures and anxieties, you must never lose your connection to the whole organization. Even as you are looking outward, you have to find ways to constantly see down to the very lowest level of the pyramid and into the most remote corner. If you don't know what is happening down there, you will make mistakes up at the pointy end.

If you don't rise that high you did not fail. Only a few can rise to the top. Most who didn't continue to make the place work. They are no less important than the guy at the top, no less dedicated; they contribute no less to the success of the organization. I don't measure your success by your rank or position, but by the contribution you are making.

Many officers I have known should not have been promoted. They were people who performed well at their previous level but whose potential for the level above had been misjudged, and they failed. A few were so overwhelmed by the responsibilities and expectations of the higher level that they fell into depression. When we are considering moving people up, their previous record is important, but at least as important is their potential to be successful at the higher level. It's not easy to judge that potential, but time and experience help.

I always evaluated candidates by what I call the 50-50 rule. It works this way. I score 50 percent for their previous record. They had to have demonstrated proficiency, but that is just the ante to get into the game. The other 50 percent is that intangible, instinctive judgment I'd picked up over the years to measure someone's potential to do even better at the next level. Though I was pretty good at it, I wasn't perfect. I frequently made mistakes. I'd missed something in my evaluation, or I'd been swayed by friendship or insufficient diligence.

As I moved up, I always kept in mind the story of the old general sitting at the officers' club bar staring into his third martini. A brand-new second lieutenant comes in and spots him. He can't resist sitting next to the general and starting up a conversation. The old gentleman patiently listens to the kid and courteously answers his questions. After a time the lieutenant gets to what he really wants to know: "How do you make general?" he asks with raw, unconcealed ambition.

"Well, son," said the general, "here's what you do. You work like a dog, you never stop studying, you train your troops hard and take care of them. You are loyal to your commander and your soldiers. You do the best you can in every mission, and you love the Army. You are ready to die for the mission and your troops. That's all you have to do."

The lieutenant replied with a soft voice, "Wow, and that's how you make general . . ."

"Naw, that's how you make first lieutenant. Just keep repeating it and let 'em see what you got," said the general, finishing off his last martini. Then he left.

When I was a young second lieutenant, I loved my job. I loved the Army. I put everything I had into doing the job well. And I was content. Nothing was promised, and I had few expectations. Count on maybe becoming a lieutenant colonel and retiring with twenty years of service at half pay, I was told. Just be thankful for anything that comes after that and thank your soldiers for making it happen. If you hit the walls of the pyramid, find satisfaction there. Be happy with that prospect. And I was.

Chapter Nine
Potential, Not Just Performance

In the Army, we are measured constantly and exhaustively. We get evaluation reports annually and every time we change jobs or our supervisor changes jobs. Our immediate superior evaluates us. So does our next higher superior, and his evaluation compares us with all our peers who serve under him. Our school performance is graded. Our spouses are silently observed. Our careers are obsessively examined and managed.

The reason is simple and obvious. We do not hire from outside. If we need a battalion commander fifteen years from now, we have to grow one now from a promising new second lieutenant. Sergeants major are not hired in from Walmart or Hertz. It takes many years to grow them from basic training recruits. I was told as a lieutenant that only one out of a hundred of us would become a general. Ah, but which one!

Performance evaluations determine that choice. They are an essential part of the promotion system. We are bended, folded, and mutilated throughout our careers.

Though necessary and useful, performance evaluations don't give the whole picture. Past performance alone does not adequately predict future performance. Sure, if past performance is mediocre or worse, satisfactory or outstanding performance in the future is extremely unlikely, and if past performance ranges from better than satisfactory to outstanding, chances are good that performance in the future will continue at that level. But it's not a sure thing.

In both the military and civilian worlds, evaluations of potential are mostly subjective, or even anecdotal. "She's going all the way." . . . "He's got General Officer potential." . . . "She's a winner, promote ahead of others." . . . "He's a water-walker." Judgments like these are based on more than performance. Leaders and bosses see qualities that separate some few from the crowd. What do they see?

For starters, they see consistently outstanding prior performance in different positions.

They see someone learning and growing intellectually, someone preparing for the next level, not just maxing out in his current job; someone who is ambitious, but not cutthroat.

They see someone tested by assignments and challenges generally given to people with more seniority and greater experience, thus indicating early that he can probably perform well not only at that higher level, but at levels above that one.

They see someone reaching outside his comfort zone to acquire skills and knowledge that are not now essential, but are useful at a higher level.

They see someone who has demonstrated strength of character, moral and physical courage, integrity, and selflessness, and who will carry those virtues to the next level.

They see someone who is confident about the next step. His ego is under control, and he is mentally prepared for the added responsibilities and burdens of higher office. It won't go to his head. He is balanced.

They see someone who enjoys the respect and confidence of his contemporaries who may soon become his juniors.

Even when someone passes this kind of evaluation with high marks, mistakes can be made.

After an officer I knew was promoted from colonel to brigadier general, an inadequacy surfaced that had not been detected earlier, and he broke under the burdens and expectations that were placed on him. One morning he committed suicide in his garage. We missed signs and portents we should have seen. He would have served successfully for many more years as a colonel, but we raised him up to a position beyond his potential.

His was an extreme but not uncommon case. Many people cannot scale up to the next level. I have known officers who asked not to be considered for promotion. They were satisfied with their place in life, realized they couldn't handle greater responsibilities, and had the courage to act accordingly. A promotion would have made them miserable.

On the other hand, we sometimes missed an officer's true potential.

There are many kinds of executive positions. Someone who can't thrive in one may perform spectacularly in another.

Colonel Dick Chilcoat, my executive assistant when I was Chairman of the Joint Chiefs of Staff, came highly recommended and I hired him sight unseen; I had never met him. Though he was only a few years younger than me, he had been passed over several times for promotion to brigadier general. Some problem earlier in his career had held him back.

Dick performed brilliantly for me, and I thought he should be promoted, but I realized that since he had missed promotion twice before, I needed to make the case that he had other talents that may not have been adequately considered by the promotion boards. I gave him a superior evaluation that pointed out another dimension of his potential—as an educator. The promotion board agreed and he was promoted to brigadier general. He went on to be promoted twice more, and rose to become commandant of the Army War College and president of the National Defense University. After retiring as a lieutenant general, he became dean of the George H. W. Bush School at Texas A&M University. He was a master educator. Dick, I'm sad to say, passed away in 2010.

At other times I've recommended someone for promotion to the next higher level with the understanding that he would go no further. This person had unique abilities that we needed one step up, but he had to stop there. No higher. His unique abilities would not be needed at a higher level, or he did not have the additional abilities the higher level required.

Needless to say, it may happen that he grows and expands his potential, or else the higher-level position may change and require his unique abilities.

The leader must understand his subordinates, an imperative that includes identifying, training, watching, mentoring, encouraging, and evaluating the next generation of his organization's leaders.

But leaders are not gods. Their understanding is never totally clear, totally accurate, totally certain. Every leader is human . . . imperfectly human. Water-walkers sometimes fail, and quiet walkers sometimes end up on top. Leaders need to watch all their subordinates; work with all of them, encourage the hotshots, but invest in the others. Always be prepared to change your mind, however firmly made up, when dealing with those infinitely faceted beings we call people.

The leader must never forget that he may end up working for one of them.

PART III

Take Care of the Troops

Chapter Ten
Trust Your People

In the early days of George W. Bush's presidency, the State Department began planning for the President's first trip out of the country, a meeting with the new president of Mexico, Vicente Fox, at Fox's ranch. There were important issues to discuss, among them immigration, border control, drugs, and trade.

In preparation for the trip, I asked President Bush to visit the State Department to be briefed on Mexico-related issues. It would be his first visit to State since becoming president, and I knew it would give my troops a boost. He readily agreed.

At my staff meeting the next morning, I explained how I wanted the briefing to be handled. The two junior Mexico desk officers were going to brief the President. The young Foreign Service officers at the boiler room, desk officer level should know more about what was happening on the ground in Mexico than anyone else. When the time came for the President to be briefed, I would merely introduce them. No senior officials would speak, no assistant secretaries or deputy assistant secretaries. The staff gave each other skeptical looks. "When would you like the rehearsal?" they asked after a pregnant pause. "When would you like to check the slides they'll use?"

"I don't want a rehearsal," I told them. "And I don't need to see slides." Frankly, I didn't want slides. No PowerPoint. The two junior officers would just sit across from the President at the conference table and tell him what they knew and what he needed to focus on and remember.

I had little concern. I had never met the two officers; I didn't even know their names, but I was sure they'd be ready. They would spend the days from now until the briefing working like dogs, consulting with their bosses and the embassy in Mexico City, reading everything they could, and getting ready for their big moment. They might lose a little sleep. They might feel more pressure and excitement than they were used to; and their spouses were doubtless calling every living relative to share the news.

In fact, the whole building was buzzing. I expected that. I wanted it.

The day came; the President and his party entered the conference room and took their places on one side of my large conference table. The table was historic. It had been used at the 1983 G7 Summit in Williamsburg, Virginia; a plaque at each place identified the head of government who had sat there.

I welcomed the President, introduced my key leaders, and then introduced the two action officers and turned them loose. Of course, I had briefed the President about my plans, and he was eager to play his part. The two officers took off, and their performance totally met my expectations. They provided the President with all he needed to know before he flew down to Mexico. The President asked penetrating questions and got solid answers. When it was over, he expressed his satisfaction, thanked everyone with a handshake and a smile, and swept out, assistants in his wake. I'm sure the two officers rushed back to their phones to call home; all their office mates must have then clustered around for a debriefing.

Here was the real payoff. Word went around the department at the speed of light: "It was great! The new Secretary trusted us. So did the President." Over the past ten years, dozens of State Department officers have reminded me of that story.

I believe that when you first take over a new outfit, start out trusting the people there unless you have real evidence not to. If you trust them, they will trust you, and those bonds will strengthen over time. They will work hard to make sure you do well. They will protect you and cover you. They will take care of you.

This isn't a fairy tale of confidence building. If the briefing had gone wrong, I would have known immediately that I had more serious problems than I had so far recognized, and that I might have to take drastic action. However, my style is not to expect trouble when I take on a new outfit. I like to go in believing that the leaders who were there before me were smart and had done their best. I'd learned long ago not to go in swinging a samurai sword like John Belushi in a *Saturday Night Live* skit. All that does is put people on guard, and make them anxious and afraid. The sword swinger is seen as an infection, and bureaucratic white corpuscles will race to attack it.

During those same early days at the State Department, I asked my principal line officials, my Assistant Secretaries, if they were reluctant to go up on Capitol Hill to deal with members of Congress. Hands went up: no one liked to go up to the Hill. I could understand that. I didn't relish it either. But I still had to do it, I told them, and it was too heavy a load for me to lift alone. I needed them to carry more of it. Their reluctance stemmed from concerns that they might say the wrong thing; get in trouble, both up there and back at the department. I told them I would make sure they knew the administration positions, and I would expect them to defend those positions. I trusted them to do so. They didn't need to check in with me beforehand—just go up there and see what the member or the committee wanted. Always approach congressional questions with a "Glad you asked!" attitude. They are the people's representatives and we are the people's servants. And if you get in trouble, we'll work together to get you out of trouble. We're a team.

There will be times when you need to take up a sword.

When the Iran-Contra scandal shook the Reagan presidency in 1986, Frank Carlucci, Howard Baker, Ken Duberstein, and I were brought into the National Security Council and the White House Chief of Staff's office to cut out the infection and stem the bleeding. We did that, and in the process we fired lots of people. But we embraced those who remained, and the new people we brought to the team worked well with those we kept on the basis of mutual trust and a commitment to making the last two years of the Reagan presidency a success. We achieved that goal.

When I first entered the Army I was sent to Fort Benning, Georgia, for my basic officer training. At the end of the course, a wise old sergeant said to me, "Well, Lieutenant Powell, you are off to a good start. You might make it in the Army. But let me tell you something about leadership. You'll know you are a good leader if your troops will follow you if just out of curiosity. The day will come when they are facing life-or-death danger, they are scared and unsure. Yeah, you've trained them and they've got the weapons and equipment to get the job done. They are curious as to how you are going to get them out of this mess and will stick with you to see."

The sergeant was not really talking about curiosity, but about trust. They will follow you because they trust you. They will follow you because they believe in you and they believe in what they have to do. So everything you do as a leader has to focus on building trust in a team. Trust among the leaders, trust among the followers, and trust between the leaders and the followers. And it begins with selfless, trusting leaders.

Chapter Eleven
Mutual Respect

Leaders have legal authority over followers. They can demand and expect obedience on the job and have the power to take action against followers who do not obey or meet expected standards of performance. They can fire them. They can dock their pay. They can demote them. In the military we have severe punishments for disobeying orders.

Obedience alone may get the job done, but it probably doesn't inspire commitment to the job. It doesn't necessarily inspire pride in the work or the product or a passion for excellence. These come when followers feel they are part of a well-led team. And this comes when they respect their leaders, and when they, in turn, believe that they are respected by their leaders. It comes when they trust their leaders, and when they believe they are trusted by their leaders. They have to know they are valued.

You may be able to run an assembly line without having the respect of the workers on the line. They meet the quota, they get paid by the hour or by piecework, and that is the deal they have with their leader. In exchange for pay they agree to become part of the machinery.

Even on factory assembly lines respect and trust between leaders and followers may inspire line workers to exceed design expectations and motivate them not to slack off.

Respect for leaders by followers can't be mandated; it must be earned. It has to be given to leaders by their followers.

You gain their respect by knowing and respecting them and through your own competence and personal example. Yet leaders must maintain a certain distance; they can't get too close. Followers want leaders who are selfless, not selfish. They want leaders who have moral and physical courage, who always do the right thing, and will risk their careers in so doing. They want leaders who are tough but fair, and never abusive. Leaders who not only are role models, but also inspire followers to be their own role models.

When such an environment exists in an organization, it hums, and you can feel it. The followers will take care of you and will see to it that you and the organization succeed. They will internalize that passion to succeed.

One miserable day in Korea in 1974 my battalion was called on to assemble in the post theater *right now* to listen to a speech from a visiting Pentagon official. With no prior notice we were expected to fill the theater in twenty minutes. The unit was spread all over the post. I complained briefly, but was told I was wasting time, get on with it!

The theater was locked. We had to knock the lock off with an ax. Troops were dragged in from all over; wanderers from other battalions got scooped up. We even dragged in a soldier on his way to the stockade and his two MP escorts. We filled the theater just in time. The Pentagon official arrived, gave a ten-minute speech on race relations, and was gone.

The bewildered troops staggered out of the theater wondering what the hell that was all about. I felt miserable and imagined the troops mumbling about military dumbness and their idiot battalion commander. As I walked to my office, one of my company first sergeants came alongside and announced cheerfully, "Hey sir, it's another great day to be a soldier."

"I don't think so," I said. "I just jerked the whole battalion around for a dog and pony show."

"Hey sir, no problem," he replied. "The troops are fine. They know you needed them there and you would never have come up with such a nutty thing. They are with you."

I brightened instantly. No recognition I ever received has meant more to me than his.

If you want to respect your followers, you have to know them. When I was starting out as a lieutenant, I was taught to learn all I could about the few dozen soldiers I was responsible for. I kept a pocket notebook with a section for each soldier, listing his name, birthdate, serial number, rifle serial number, family members, hometown, education, specialty, date of rank, and my initial and subsequent observations about his performance, conduct, appearance, ambitions, strengths, and weaknesses.

As I moved up into larger organizations, and was no longer in daily contact with all my followers, a small notebook no longer worked. I used direct reports from my staff to keep me informed about everybody under me. I didn't just want formal reporting and report cards. I wanted naked truth. Who was being naughty, who was being nice? Who was inspiring his followers? Who had a family problem or an emotional problem? I tried hard to find out what people didn't want me to know. I needed to know if such things affected their performance, and I needed to make sure they were doing what I wanted done and expected them to do. Were they operating in harmony with me?

During my four years as Secretary of State I tried to get to know each of our ambassadors, the President's principal representatives to other countries. I made a point of swearing in every one of them at a formal ceremony with a large audience of family and friends (I was unable to do this only when I was out of the country). I presided at 145 such ceremonies. I considered it a laying on of hands, cementing a bond of trust and respect between the ambassador and me. I made it clear to all the ambassadors that they were free to call me directly anytime, seven days a week, at the office or at home. I am never too busy for you.

After they took their posts, I closely monitored their performance. My regional assistant secretaries of state knew to let me know how they were doing, especially if trouble was brewing. In turn, I kept the President informed about their performance. On three occasions ambassadors had to be quietly removed before formal channels woke up to the problem, because of information I had received through informal channels.

Another attribute necessary to gaining respect is competence. If you don't know your job and can't do it well, there is no reason why followers should respect you. I am sure that just as I was writing a page about each of my soldiers those many years ago, they were all in their own way writing their own page in their mental notebook about me. Does the lieutenant look sharp? Can he keep up with us at PT? Can he shoot a rifle or drive a tank almost as well as we do? Does he take care of us? Does he listen to our problems? Does he ever try to con us? Is he tough or soft? Does he trash talk his boss or other lieutenants? Does he protect us? Does he accept blame and share credit? Do we like him?

I have no doubt that my assistant secretaries of state had even longer lists that they constantly exchanged with each other.

A leader needs to know his followers, and he must be competent; but he is also an individual; he needs to preserve a zone of privacy, a place for himself that his followers can't enter. They need to be kept at a distance. There is an old expression attributed to Aesop: "Familiarity breeds contempt." It might be better said that too much familiarity brings everyone down to the same level. The leader is with the troops, but above them. He should always maintain an aura of unpredictable mystery.

Though every leader wants his followers to like him, and followers want to like their leader, liking is not necessary. It helps the organization run more smoothly. But if respect is lacking, the organization will probably run badly. Liking has to come from respect, not from the leader trying to be a nice guy or a buddy to the followers. They don't need you to be easy on them.

A certain air of separateness is essential. Followers are not your buddies; they are your followers, your subordinates. If you aren't different from them, if you don't provide them with what they can't do for themselves, then they don't need you.

I've often heard blowhard leaders boast, "My outfit is so good, it could function well without me." Hmm, then why do they need you? The leader is always above, but never beyond, the followers. So a leader can socialize with his followers, but not to the point of hanging out with them. Friendliness is fine, short of familiarity. Never let a follower mistake liberty for license.

Finally, real leadership and unfailing respect are a retail issue. They happen on the ground, where the troops are. They don't come out of directives from on high.

One night back in the 1970s, I was driving home to my quarters at Fort Campbell, Kentucky, where I had commanded the 2nd Brigade of the 101st Airborne Division for about a year, when I saw in the dark a soldier walking along the road heading for the gate. He probably lived with his wife in the trailer park just outside the gate. I stopped and offered him a ride.

"Why are you going home so late?" I asked him as we drove along.

"My buddies and I've been working hard to get ready for an inspector general inspection coming up," he answered. Then he looked at me. "Sir, who are you?" he asked.

"I'm your brigade commander," I told him, taken aback.

"How long have you been in command?" he asked.

"Over a year," I said.

"Is it a good job?" he asked.

"Yes, great," I replied. Jeez, after a year of being all over the brigade area, here is a soldier who doesn't recognize me. Something's wrong.

"How do you think you guys will do in the inspection?" I then asked.

"We'll do great," he answered. "We've been working hard for weeks, and my captain, lieutenants, and sergeants have been pushing us. They've been telling us how important the inspection is; they've been working just as hard as we have." Then he said simply, "We're not going to let them down." Something's right.

I was the brigade commander, but it was his buddies, sergeants, and officers who were his family, who trained him and provided for him; who took care of him. And, in turn, that care and family feeling would flow up, and they would take care of me. Mission accomplishment starts at the bottom.

What moved me the most was his saying, "We're not going to let them down." As a leader, you will never receive a better compliment from your followers. You will never have a better report card showing how you are doing. You've created a winning team. A team that rests on a solid foundation of mutual trust and respect. They will never let you down as long as you never let them down. The troops will always get it done and take care of you. Make sure that every hour of the day you are taking care of them.

Chapter Twelve
We're Mammals

I love to watch nature channels on TV, and I especially love animal shows about our fellow mammals. I will watch any show about mammals, but shows about lions are my favorite.

A mother lion has a litter of cubs. They are kept in a den for several weeks, until their eyes are fully open and they have fully bonded with their mother. Papa lion is out there somewhere doing king of beast things; the mother does the nurturing. After a couple of months the cubs are allowed to explore their surroundings. Momma, watching, keeps them within strict boundaries close to home. If they stray outside her box she calls or drags them back inside.

With time the cubs grow and they master the territory inside the mother's box. The boundaries of the box grow larger. Later, she takes them roaming with her outside the box and teaches them how to hunt, but continues to provide food and structure. They are still learning.

At about two years old, the cubs are allowed to drift off on their own. But before that step, as young cubs they learn the collective wisdom of a thousand generations by observing their parents and relatives. They learn how to survive as a lion and what it is to be a lion. They learn the proper way to behave in the group. They learn to hunt by watching and following the adults, not by being briefed by PowerPoint. They learn what is expected of them the same way. The adults guide them and don't let them get ahead of their age and experience level. To be abandoned by parents, especially by mom, is usually certain death.

I also love elephants.

I will never forget a *National Geographic* elephant documentary I saw years ago (it's still shown frequently). Several adolescent male elephants were removed from their herd and transferred to an isolated reserve where there were no other elephants. All hell broke loose. Within weeks the adolescents started acting erratically, becoming generally belligerent, even attacking and killing rhinos, who are not natural enemies or competitors. Their testosterone levels were out of control.

The park rangers began to worry that they would have to destroy the adolescents. Wiser heads prevailed. They imported several adult male elephants. The adult males asserted themselves, and almost immediately, in the presence of the adults, the juvenile delinquents settled down and learned that elephants don't kill rhinos. Even testosterone levels went down as centuries of elephant experience was conveyed to the juvenile delinquents.

I don't speak elephant, but I can imagine tough but loving conversations like: "Hey, dude, elephants don't do that." Or: "Don't make me come over there and slap you upside the head with my trunk."

The best advice I ever received did not come in the form of words or aphorisms. I got it from watching my parents. Yes, they lectured me, passed down the usual old wives' tales and collected family wisdom of several generations. I am sure I internalized and benefited from all that. But the most valuable advice I got was from their example, how they lived their lives. Children may or may not listen to what parents are telling them, but they are always watching what their parents do. "Do unto others" is timeless universal advice. Children will learn and live by that injunction forever if they see their parents reaching out to help others in need. If parents respect each other and create a climate of love in the home, children will see the value of that environment and will try to replicate it as they grow up.

Are we the only mammals dumb enough to forget where we came from, what we are, or what we can't do without if we are going to live and grow well? Are we beginning to lose our understanding of the importance of tribes? I'm afraid the answer is yes.

We don't live on our own. Tens of thousands of years ago, when humans were emerging on the African savannas, our ancestors did not survive as solitaries; they survived and worked together in bands. They learned and grew and optimized their capabilities in bands and tribes, not on their own. That remains true.

Adults need to pass on all our generations of experience. Children need to know that their herd is their family, always there for them. They belong to a tribe. A tribe that will protect and guide them. They should know all this and have that tribal support when they start school.

Education begins the moment a baby hears her mother's voice and realizes the voice is her mother's. Babies need nurturing and structure. They need boxes to be safe in and in which to grow and learn, with parents and families watching, correcting, and above all, loving them. Children need to be taught early in life what is expected of them and how they must never shame their family. They must be taught to mind their adults. If a kid isn't spoken to properly, read to, taught numbers, colors, time, how to behave, how to tie his shoelaces, play nice, share, respect others, and know the difference between right and wrong, he will be miles behind by the time he reaches the second grade; it takes that long for the kid to know he's behind and to start acting behind. He will from then on have trouble keeping up with other children—an all too familiar problem in our society.

But it can be fixed. Early childhood programs like Head Start and after-school programs, as well as inspired teachers, coaches, ministers, successful people willing to mentor—all of these interventions can keep kids from joining a bad, failing tribe.

Above all, kids must be taught that they are ultimately responsible for what they achieve or fail to achieve. Overcoming obstacles is a part of life.

There is nothing complicated about this. Without the example I saw in my home and in my extended family, I wouldn't have succeeded in life. They always let their "light so shine before them that all knew of their good works."

I once watched a television piece about Arrupe Jesuit High School in Denver, which serves poor inner-city neighborhoods. All seventy-one graduates that year were going to college. An interview focused on a student named Jose, who was the valedictorian. He was the first member of his family to finish high school.

"How was that possible?" the interviewer asked.

"I was never, ever given the opportunity to fail," Jose answered simply. "People kept pushing me. They picked me up when I fell. They believed in me. If they felt that way about me, I had to feel that way about me." And then he added, "I have changed the history of my family."

Yes, he has. He will achieve success after college and in time will raise children who will never be given the opportunity to fail and who will follow in his footsteps.

The Army is neither a tribe, a herd, nor a family, but it's not completely different from them, either. They all, for example, shape young members into the group in much the same way. Military organizations, naturally, require a far higher level of discipline than nonmilitary ones. You mold young men and women into soldiers only with order and structure. In the Army, the people who fit the raw recruits into its ordered and structured mold are the sergeants—the experienced elders who model for recruits the way you have to live and act in the Army.

The first thing a recruit learns is how to stand at attention in formation—an efficient way to put him into a structured box and to move numbers of them about efficiently. It also teaches conformity. If the drill sergeant says "right face" and Joe Six-pack goes left, his whole platoon looks goofy, and it's his fault. Faults are immediately recognized and have immediate consequences.

The new recruit shares an identical haircut with his buddies and wears the same clothes. No bling, no distinction.

Drill sergeants cut recruits no slack, work them to exhaustion, and allow only three answers to any question: "Yes, Sergeant. No, Sergeant. No excuse, Sergeant." As in, "I don't care how many times you cut your face, you need a shave."

"No excuse, Sergeant."

Try something like that with a sixteen-year-old in your house.

So it goes for a number of weeks. The recruits come to resent, nay, detest, the drill sergeant. Then something fascinating happens. They start to learn things. By the time basic training is over, they don't hate their drill sergeants, as tough as they have been. Instead, they want to please them. They would just as soon never see them again after graduation, but they will never forget them. I once asked the late senator Ted Kennedy if he remembered his drill sergeant. Yes, he certainly did; he regaled me with stories for half an hour.

In 1989, during my time as Commander of the Army Forces Command, I was taken on a tour of the weapons systems at the Army's Air Defense Artillery School at Fort Bliss, Texas.

At the Patriot missile system display, a young Hispanic soldier (he looked no older than nineteen) was waiting beside the control system van to brief me. We chatted for a moment before he started his presentation. I was curious about where he was from; his slight accent gave me no clues. Lo and behold, he was a public high school graduate from New York City like me—a New York street kid. He had been in the Army about eighteen months.

When he swung into his presentation, he flawlessly described every component and function of the control system—the range of the radar and the missiles, the number of targets it could track and engage, and the electronics in the van, at which point I got lost in the technical details. As a general, my facial expression had to display total comprehension, but my actual incomprehension just reminded me of why I dropped out of engineering at the City College of New York.

How did this street kid have all this complex information at his fingertips? I wondered. Did he really understand what he was talking about? Or was he just going on rote memory? I interrupted him a couple of times with questions to see if he could pick up his flow again. He didn't lose a beat. He really knew it.

This didn't surprise me. I'd observed scenes like this hundreds of times. I glanced around to confirm my instinct. Sure enough, a sergeant was standing just around the corner of the van, barely visible, but close enough to hear everything the GI said. As the soldier talked, the sergeant was mouthing his words. When I broke in with questions, the sergeant froze; the GI was on his own to handle the answers. When the GI fielded my questions perfectly, the sergeant relaxed.

He was the soldier's sergeant, his boss, the one who had trained him, drilled him, tested him, and expected the best from him. He was the one who had filled this New York street kid with the confidence to stand there and belt it out to a four-star general and a bunch of other senior officers. The soldier had someone who believed in him. The soldier would not let him down.

When the briefing was finished, I congratulated the young man, wished him well, and proceeded toward the next display station. I've been around the Army a long time. When I was maybe twenty feet from the van, I glanced over my shoulder, knowing what I would see. The sergeant was high-fiving the kid, and all his crew buddies where giving him a "Hoo, hah."

All followers need to feel they belong to a team, a tribe, a band. Leaders are leaders because they pass on the generations of experience they have amassed. They give purpose to the team, give it structure, hold it to standards, nurse and nurture the team, slap it upside the head, as needed, and above all give the followers someone to look up to.

Chapter Thirteen
Never Walk Past a Mistake

This is one of the first lessons drilled into young military leaders.

To put it another way: make on-the-spot corrections.

This serves a number of purposes. First, and most obviously, correcting a mistake shows attention to detail and reinforces standards within an organization. Thus a young second lieutenant will always correct a soldier who fails to salute when he is passing by or who is wearing his insignia an inch off center. Tolerance of little mistakes and oversights creates an environment that will tolerate bigger and ultimately catastrophic mistakes.

Second, it teaches aspiring leaders to have the moral courage to speak out when standards are not being met. You never look the other way and pretend you didn't see it just to avoid a confrontation or to be seen as petty.

Third, it shows the followers that you care about them, the unit, and its mission. If a follower knows that he has just made a mistake and gotten away with it, he loses confidence in the competence of the leader and has less respect for him.

Fourth, you set the example for all of your subordinate leaders to act in the same manner. High standards and mutual respect will flow up and down the organization.

Fifth, it keeps mistakes and screw-ups from moving to another level or, even worse, propagating. Take care of it now. Don't assume somebody will take care of it later . . . even if it's their responsibility.

Attention to detail and on-the-spot corrections need not devolve into silliness. A group of soldiers just in from the field, dirty and tired, should not be nailed for being dirty or a little lax. Common sense should prevail.

I have found that corrections done in a firm and fair manner with an explanation are appreciated, not resented. Always try to turn the encounter into a mutually positive learning experience.

These truths are known to every good classroom teacher, every good coach, every good violin teacher, every good parent, and every good construction foreman. Mistakes that have become deeply rooted habits—in a batter's stance, in a violinist's fingering, in a child's table manners, in a roofer's roofing skills—drive teachers, coaches, foremen, and parents nuts. You have to catch them all early, and properly train the correct actions, skills, and behaviors. Leaders who do not have the guts to immediately correct minor errors or shortcomings cannot be counted on to have the guts to deal with the big things.

Chapter Fourteen
The Guys in the Field Are Right and the Staff Is Wrong

Whenever I took command of a unit, I announced early on that my bias was toward the guys in the field; I took their word as ground truth. Until I was persuaded otherwise, my staff must be wrong. This did not make my staff happy, but that was good.

My bias toward the guys in the field may sound unreasonable, but here's how it worked for me. First, it let my staff know that our clients were the leaders on the line and their troops. My staff didn't work for me. My staff worked for them. Problem solving went both down and up. Once every staff member realized that any field commander could drop a dime on them to me, they worked like the devil to solve field problems. The staff realized they couldn't make me happy unless the line was happy.

Ah, here's the flip side. When one of my commanders complained about some staff screw-up, he knew I would look into it totally convinced that he was right. If I found out that the commander was wrong and my staff was right, and he should have known it, then it was time for a come-to-Jesus. Such actions did not endear him to me.

After a few weeks, everyone on my staff got it. "Jeez, we're in this together," they'd tell the field guys. "Please, let's both of us work on your problems before you tell the CO the next time he's down there for coffee. By the way, your monthly maintenance report was a mess and came in late. How can we help you fix it before next month? The old man is kind of nutty about this stuff and we need to protect each other."

Over my many long years of experience, the line was right about 70 percent of the time.

Chapter Fifteen
It Takes All Kinds

When I commanded the 2nd Brigade of the 101st Airborne Division, I had several gifted commanders with radically different personalities. I could tell one of them to go take a hill, and it was done. I could tell another to go take a hill and he immediately asked questions: "When? How? What other support can you give me? Do I have the priority on resources? Then what do I do?"

Both commanders would take the hill and accomplish the mission. Who was the better commander? The can-do commander was exciting and admirable, but he sometimes charged off without asking basic questions. He could get in trouble quickly. He did not always capture all my guidance and the larger picture of the battlefield. The other commander could annoy me with his pestering questions, but he often came up with the more skillful plan and the more careful execution.

My job was to get the best out of both and complement their strengths and shortcomings. I would often question the can-do commander to make sure he understood the answers to the questions he didn't ask. His aggressiveness and can-do attitude needed to be monitored and controlled. And I would often lose patience with the pesterer, cut him off, and toss him into the action.

You will seldom get a perfectly matched set of subordinates. They are not clones, and even clones aren't identical. As long as they understand what I want and are in harmony with me, I can manage their differences. Within the range of my personality, experience, tolerance, and expectations I can work with practically any combination of subordinates—as long as they can do the job.

Now and again it turns out that a subordinate is not in harmony with me, and I have to relieve him. This is never easy. It can be especially difficult when a subordinate has done nothing specifically wrong that warrants relief. During my assignment to the 101st, I had to relieve a commander for that hard-to-pin-down reason. He'd so far had a successful career; he'd done nothing specifically wrong that would have demanded his relief; but I never sensed that I had him, that he was in my space. He performed well enough to be seen as competent, but that was not enough. He wasn't leading to my satisfaction. He executed my instructions, but only marginally and without passion and intensity. He went through the motions with minimal enthusiasm and commitment. His unit reacted in like manner. He didn't inspire or fire up his troops. I was dragging a weight behind me.

His personality made him a good manager, but not a leader who made a difference. Everyone could see it, and I had to let him go. It was difficult, but I hadn't acted precipitously or in an arbitrary manner. I had tried counseling him, but that didn't help. He knew my concerns and lack of satisfaction directly from me.

When I told him that I had to relieve him, I made it clear that I was quite sure he could be successful in another job in a different capacity. He was crestfallen, but the needs of the unit came first.

During my transition to Secretary of State, I recruited an old friend as a speechwriter. He was one of the best speechwriters I had ever met; he had worked at the very highest levels of government; and I had worked with him before and knew his style. It had all the elements of a brilliant choice.

It didn't last long. He kept trying to cram his thoughts into my words, rather than use his skills to enhance my thoughts and words. We had a heart-to-heart one evening, and he quietly found a senior official to work for in another department. I wished him all the best, but my game, my ball.

What do I look for in subordinates? The usual qualities: competence, intelligence, character, moral and physical courage, toughness with empathy, ability to inspire, and loyalty. Beyond that, I want subordinates who will argue with me and execute my decisions with total loyalty, as if the decision were originally their own. Past performance is examined closely, but I try to sense future potential. I want imaginative and creative folks with ideas and the ability to anticipate. I treasure the person who sees a problem before I do and does something about it before I even know it exists. I treasure the person who sees opportunity before anyone else and smells risks and threats early.

I also look for people who will fit in with me and with my team. When I commanded a battalion in Korea, my brigade commander asked me to give a company command to a brilliant young captain on the brigade staff. He was an exceptional officer—extremely talented—but he'd let his brilliance go to his head. He had trouble getting along with other captains. His personal behavior left something to be desired and was the subject of much gossip in the brigade. He was good, and I could have managed him, but he would not have fit in on my team.

I suggested to my boss that the captain be assigned somewhere else. He was. He got off to a solid, though too self-promotional start. But after a few months his personal behavior problems publicly manifested themselves, resulting in an ugly inspector general investigation and his relief.

On some occasions, I've passed on people I probably should have hired. On other occasions, I have stuck with people I should have let go. And many times, people have strengths you need, even when that means you have to put up with weaknesses that you forever have to cover. In selecting people you just hope you bat over .500.

Because you have to have spice in the stew, I also look for characters. An organization is invigorated when a handful of slightly felonious, offbeat eccentrics are on the team. Some of my most memorable experiences and good ideas have come from folks who get out of the box and have fun. Guys like Tiger Honeycutt. Brigadier General Weldon "Tiger" Honeycutt was my immediate boss in the 101st Airborne. A heavily decorated hero, he never flinched from taking on anybody or anything.

One weekend the senior leadership of the division was convened to participate in a two-day "Organizational

Effectiveness" seminar, run by a civilian academic facilitator. His first instruction was to list on charts our goals and objectives, after which we would discuss our feelings about them. When he finished his opening presentation, Tiger raised his hand. "How much are we paying this son of a bitch?" he asked. Tiger was excused from the course.

Every morning, I gave him my standard greeting, "Good morning, sir, how are you?"

His standard reply: "A helluva lot better than you. I'm a friggin' general and you ain't."

Guys like Tiger are the spice every organization needs.

When I'm choosing people, I try to support my strengths and fill in my weaknesses. I want people around me who are better than I am in areas where I am not comfortable. I want folks who are smarter than I am, but who neither know it nor show it.

In selecting a deputy, I always want someone who is tougher and nastier than I can be. I'm the good guy and chaplain. He is the disciplinarian and enforcer. Major Sonny Tucker was my executive officer in the 2nd Brigade, 101st Airborne Division. His office was right next to mine; I heard just about everything that went on in there. When someone had displeased me, all I had to do was let Sonny know. Later that day I could hear him through the wall. "C'mere, boy, you made my colonel unhappy; and when he is unhappy, I am pissed. And now I am going to eat your lunch." After Sonny retired he became a minister.

I always fully empower my deputy to act for me.

Early in my tenure as Secretary of State, my deputy, Rich Armitage, was given a document to sign while I was on a trip. The staff put the title "Acting Secretary of State" under his signature. "That's not necessary," I told them. "I'm always available through the miracle of modern communications. We'll never need an 'Acting Secretary' while I'm Secretary."

The staff was confused about what they should do when I was on the other side of the world. The answer was simple: Rich could sign as Deputy Secretary of State. His signature was as valid as mine. In the very few instances where the law required my signature, I'd sign. Those were the only exceptions.

The point was, Rich had my total trust and I had his. The staff tried to write this arrangement up in a regulation. I told them not to bother; they'd soon see how it worked. There was never a problem.

Do I look for good managers or good leaders? Let us bury that old distinction. Good managers are good leaders, and good leaders are good managers. But great leaders have a special touch that separates them from managers. Good management gets 100 percent of a team's designed capability. Great leaders seek a higher ground. They take their followers to 110, 120, 150 percent of what anyone thought was possible. Great leaders do not just motivate followers; they inspire them. The followers are turned on by their leaders.

Superior leaders also tend to be superior managers. They are rare gems. Always be looking for the person with the potential to give you 150 percent.

In the early 1980s, we were working hard to see if we could use simulators to make unit training cheaper and more effective. I was a brigadier general in the 4th Infantry Division at Fort Carson, Colorado, when we received for testing new tank gunnery simulators.

Tankers love to race cross-country and shoot their main gun. It is how they train to win in battle. But there are downsides. Not the least of these is the cost of the shells they fire, up to $1,000 each in 1981 dollars. For that reason, each tank crew got an allocation of only ninety shells per year.

This is where the simulators came in. Could the same level of proficiency be achieved using simulators? We were instructed to test our new simulators to find that out.

The crews were put in an isolated booth replicating the inside of a tank turret. The terrain rolled by on a screen; enemy tanks popped up; and the crews engaged them with an electronic gun.

We selected two tank battalions to test the concept. One got the full ninety rounds per tank and did not train on the simulator. The other battalion got only fifty rounds, but had hours of training per crew on the simulator. We then reversed the battalions and repeated the trials.

In the first trials, the battalion with no simulator time scored better. But then when we put the same battalion on simulator training, they scored higher again. We cut fifty rounds to forty rounds, and the same battalion won. We reversed the process again; the same battalion kept winning. The analysts were bewildered.

The answer was simple. The difference was the battalion commander. He was determined to do better, no matter what we threw at him. He drilled that into his soldiers. Every night they worked on it. Every man was determined to do his best and to hell with the analysts. The other battalion was a good battalion, but the commander didn't have those extra qualities of hunger, competitiveness, drive, passion, and imagination that his buddy did and that infected his whole unit.

I don't want to carry this lesson too far. Simulators are great for training, and we do a lot with them today that you couldn't dream of thirty years ago. Yet we can never have enough battalion commanders like the one who kept winning no matter what.

On the whole, I like people who work hard, have a purpose, inspire folks, spend time with their family, have fun, and aren't busy bastards. I like a happy team. I work hard to make sure my followers work hard, and I work hard to make sure they enjoy their work. That can only come from believing in what they are doing and feeling they have been prepared and equipped to get the work done.

I set high but not impossible standards. Mine are achievable with maximum effort.

I do not like to see an atmosphere of fear in an organization, where shouting, screaming, and abuse of subordinates are common. You're probably saying, "Well, who does?" You'd be surprised. I have worked in fear- and abuse-filled organizations and have seen a lot more. Their leaders were at bottom insecure bullies who substituted Sturm und Drang for leadership. I have never known any leader who got the best out of his people that way.

"What is a leader?" people ask me.

My simple answer: "Someone unafraid to take charge. Someone people respond to and are willing to follow."

I believe that leaders must be born with a natural connection and affinity to others, which then must be encouraged and developed by parents and teachers and molded by training, experience, and mentoring. You can learn to be a better leader. And you can also waste your natural talents by ceasing to learn and grow.

PART IV

Fast Times in the Digital World

Chapter Sixteen
Brainware

Diplomats live on information. It's the coin of the diplomatic realm. The flow in and out of an embassy is enormous. Limit or interrupt the flow and an embassy is a beached whale.

Since the earliest days of the nation, our diplomats around the world gathered information about the country they were assigned to, sent it to Washington, and in turn received information and directions from Washington to convey to that country's leaders.

In the early days dispatches were handwritten and sent by whatever means available, mostly stagecoaches and sailing ships. As communications technology advanced, and trains, automobiles, steamships, telegraph, and undersea cables brought us into the twentieth century, the State Department adapted . . . and continued to adapt as the twentieth and early twenty-first century brought us radio, telephones, fax machines, satellites, video links, and many other techniques, culminating in the Internet revolution, with all its many applications.

Adapting to new technologies brings with it many challenges. It's one thing to install new hardware and software. It's far more difficult to change people's *brainware.* Even if your people have at their fingertips the latest computers, Internet connections, smartphones, iPads, and data-crunching systems, their heads and hearts may remain in the twentieth century . . . or earlier.

When I became Secretary of State in 2001, I walked in as a born analog information junkie working hard to become digital. I realized that ancient temporal, spatial, political, cultural, and social barriers to getting, sending, and sharing information had been knocked down or massively penetrated. Information, capital, risk, opportunity, and social connections were speeding around the world at the speed of light. Before joining the State Department, I had served for several years on the board of AOL. I had a pretty good idea of what was possible. And I had learned a lot from my son, Mike, who was chairman of the Federal Communications Commission from 2001 to 2005. But not as much as I learned from just watching my digitally hardwired-from-birth grandkids. Abby and

PJ, the two youngest, then four and two, once started screaming in the back of the car their aunt Linda was driving, "Auntie Linda, Auntie Linda, you didn't turn on the GPS. We won't know where we're going. We'll get lost."

When I became Secretary, I was anxious to see how the department had been keeping up with the information technology revolution. The picture I saw was unsatisfactory. We had many generations of computers and incompatible systems—including a large number of antique Wang desktop computers, running legacy programs. (Wang had gone into bankruptcy eight years earlier, a geologic era in the technology world.)

Most State Department computers were not Internet capable. Of the desks that offered some kind of connectivity, many had two computers, one connected to the internal State Department unclassified network and the other to the internal classified network. Internet access was normally provided by at most a couple of dedicated computers shared by an entire office or even an entire floor. Other challenges included lack of security certifications and firewalls, little or no budget planning for information technology, and decentralization of systems throughout the department—all resulting in a waste of money, space, and people.

We had not invested money in the people and equipment needed to keep up with the rapidly evolving technology.

The other big problem I found was a long-standing controversy between the State Department and the Central Intelligence Agency over who should have responsibility for designing and providing the communications pipes into many of our more than 250 diplomatic and consular posts worldwide. Our embassies and consulates are not just staffed by State Department personnel; they are aggregations of people from many different U.S. government agencies. They all need information access.

Congress was not happy with the way the system was being managed. State and the CIA had tried to appease critics and different constituencies by creating a bureaucratic kluge. Responsibility for the communication pipes was switched yearly between State and the CIA. Congress reacted to this absurdity by creating a new office within the Office of Management and Budget to manage the system. Ugh. My only responsibility would be to pick the leader of the OMB office, who, according to congressional dictate, had to come from the private sector. Luckily, when I became Secretary the congressional solution had not yet been implemented. We still had a small window to find a more realistic solution to the problem.

Under the leadership of my Undersecretary of Management, Grant Green, we started to fix things. First, we pleaded with congressional leaders to hold off on creating the OMB office until we had time to review the situation.

Next we made a deal with CIA director George Tenet to conduct a study to determine who was best suited to contract for, install, and maintain embassy broadband pipes, State or the CIA. The study gave the nod to the CIA, and I took my staff out of the game, but not until securing an agreement that I would set the communications requirements, approve the CIA's candidate to manage the pipes, and provide an annual report card on the manager. Tenet agreed, and we signed a treaty. Within a year, our communications capacity had significantly increased, costs had dropped, and Congress got rid of the OMB law. Soon thereafter, State determined that much of our communications traffic could be sent securely over commercial Internet circuits, giving us an even more reliable and less expensive capability.

Meanwhile, we worked on our hardware needs. After a series of false starts with private contractors, we asked our staff to determine our computer requirements. They concluded that we needed more than 44,000 new computers, and we persuaded Congress to fund them.

Soon we had placed an Internet-connected computer on every desk in every embassy and every office in the department; every user had access to both the State systems and the public Internet. We accomplished this installation in less than two years. The last embassy we brought up to date was in Gabon; they complained about being last.

At the same time, we budgeted to avoid obsolescence. Four years down the road we would start replacing our by then out-of-date systems. We also developed a new messaging capability to move us from the world of telegraphic communication and diplomatic cables to email-based systems. We even allowed mobile devices to access our office systems. In short order, we moved from 1945 to 2001. The system is even better today.

This is another example of "taking care of the troops." You have to give your troops the tools they need to get their jobs done, or they will have no reason to believe in you or take seriously your missions and goals.

Because the State Department lives on the information flow in and out of embassies, I performed this little test whenever I visited an embassy: I'd dart into the first open office I could find (sometimes it was the ambassador's office). If the computer was on, I'd try to get into my private email account. If I could, they passed. Their network pipes were working, and they were using their computers and the Internet.

Bringing in new hardware and software was complex and difficult, but most of the problems involved were practical and functional. Permanently changing brainware was a far greater challenge. I was determined to revolutionize the way our people thought and worked. We had to persuade the entire State Department that we were now in a transactional, not a lunar, world. We no longer lived a time-bound existence where our work and actions are measured by clocks and the passage of days. Computers and email have eliminated physical, geographic, calendar, and clock constraints to communication. Diplomatic messages no longer travel by riders on horseback, or by couriers on trains, ships, or planes.

The leader starts to change institutional brainware by setting the example and changing his own.

To complement the official State Department computer in my office, I installed a laptop computer on a private line. My personal email account on the laptop allowed me direct access to anyone online. I started shooting emails to my principal assistants, to individual ambassadors, and increasingly to my foreign-minister colleagues who like me were trying to bring their ministries into the 186,000-miles-per-second world.

State maintains on its website background notes on every country in the world. The notes are put together by the embassies, but monitored and updated by department country and regional experts and by our public affairs office. Every few weeks, I checked the background notes list, which showed the date when each note had last been updated. More than once I found notes that hadn't been updated in over a year. I fussed at the staff constantly to keep all our data current.

"But Mr. Secretary, we update quarterly," my public affairs assistant secretary said, defensively and unwisely, one morning at staff meeting.

"Don't tell me we only update our website once a quarter," I said. "Walmart updates their entire information system whenever there's a transaction at a Walmart checkout counter. If I wake up and see on television that a foreign leader has died and his replacement has been announced, I want that reflected on our website background note for that country by the time I get to the office. We may not always be able to beat Wikipedia or Google, but let's try."

Years ago I gave a speech to a large meeting of Walmart store managers and senior leadership. As I waited backstage before I spoke, the crowd was whipped up by corporate leaders to football-rally emotion. There came a huge cheer, followed by shouts and congratulations. I asked my host what was going on.

"They've just announced the latest sales report," he told me.

Naïvely, I asked, "For the week, the month, or the quarter?"

"No, for yesterday," he answered. "I could give it to you for this morning, if you like."

I was surprised, but not shocked. I had seen it coming. Even before Google, Amazon, and the explosion of the Internet, big-box stores and supermarkets had realized that the technology revolution and the power and speed of information allowed them to move from a lunar world of calendar periods to a world of transactions.

Each transaction is flashed vertically and horizontally throughout the organization and triggers all kinds of actions. Inventory levels go down, profit is calculated, reorder formulas kick in, wholesalers and manufacturers are informed, replacement goods are assembled and loaded onto trucks, computers make projections. All of this happens in real time.

I set out to embed the same kind of mind-set in the State Department. Major change only works when followers realize your change has made their lives better and improved their productivity and performance.

You only know you've succeeded in implementing change when your followers believe in the change and will pass their belief on to the next generation of followers. Real change has to outlive the change agents.

I never stopped pressing our people to increase their email use and update our databases with each transaction and not at the dictates of arbitrary calendar dates. Though I am long gone from the department, whenever I travel to a foreign country I send our ambassador a courtesy email from my personal account to let him know I am passing through and will be available for calls on leaders, as appropriate. I am proud to say I get very quick responses. Embassy desktops and laptops are not being used as paperweights.

In spite of my best efforts, I could never persuade my dear friend Igor Ivanov, then foreign minister of the Russian Federation, to get online. Igor's intransigence gave me an irresistible opportunity to score points with Igor and my staff.

One day Igor called me from Moscow complaining about our UN delegation's objections to a draft resolution his delegation was presenting in New York. Our delegation believed their resolution was inconsistent with a resolution passed by the UN some years earlier. He said we were totally wrong. I was unfamiliar with the earlier resolution, and didn't have a clue what he was talking about. While I kept him talking, I pulled up the Google search box on my new computer and typed in the number of the earlier resolution. It came up in about a second. I let Igor ramble for a moment before interrupting him. "Igor, I am not sure you are right. If my memory serves me correctly, paragraph 2B(1) of that resolution suggests that you have misunderstood it."

Silence. "Colin, are you sure?"

"Well, Igor," I said, as I stared at the resolution text on my screen, "I can't be positive, but perhaps you should have your staff take a look at it again." It took his staff several hours to pull it all up. I was right; he was wrong. I loved it.

I never did win Igor over to computers. When he was in Washington, he often came to dinner. He always came with a gift.

Because Igor dressed exceptionally well, with a preference for blue Hermès ties, my gifts to him became Hermès ties. He was impressed. It must be lots of bother to get them, he told me. I took him downstairs to my home office and introduced him to the magic of online shopping. He watched skeptically as I ordered a Hermès tie for myself. It took about a minute.

He walked away shaking his head and muttering, "Nyet, nyet."

Igor was no technophobe. He was a grandmaster of the other revolutionary technology of our time—the cell phone. We conducted some of the most important conversations I've ever had on cell phones in strange locations two continents and nine time zones away from Washington.

And there were no emails to be subpoenaed, discovered, or WikiLeaked. Hmmm, maybe Igor knew something I didn't.

But even Igor couldn't avoid the twenty-first century forever. Now that he has left office and is enjoying private life and successful business activities, Igor has caught up with other technologies and we email each other.

Chapter Seventeen
Tell Me What You Know

You can't make good decisions unless you have good information and can separate facts from opinion and speculation.

I have always been a glutton for information. I wanted an overflowing in-box, lots of people dropping in to chat, constant phone calls from the staff or trusted agents telling me what they heard and saw. Over the years I learned to read quickly to get to the essence of a paper; tossing aside filler, unnecessary adjectives and adverbs, puffery, and snake oil arguments. I took the same approach listening to oral presentations: "Just the facts, ma'am, just the facts," an expression made iconic by Sergeant Joe Friday, the LAPD detective on the 1950s and '60s television show *Dragnet*.

Facts are verified information that is then presented as objective reality. The rub here is the verified part. How do you verify verified? Facts are slippery, and so is verification. Today's verification may not be tomorrow's. It turns out that facts may not really be facts; they can change as the verification changes; they may only tell part of the story, not the whole story; or they may be so qualified by verifiers that they're empty of information.

I've seen apparently verified facts go whoosh in the cold light of day. On March 19, 2003, the night before we launched the Gulf War, we were in the Oval Office receiving an eyes-on report from spies that Saddam Hussein was at Dora Farms, one of his palatial estates in Baghdad, which opened up the possibility that a successful attack there would decapitate the government. We bombed the place. The spies then reported they were sure they saw Hussein's body being brought out. All wrong.

In Somalia in 1993, we were searching everywhere for the dictator Mohamed Aidid. Spies kept reporting that they had him located, but he was always gone by the time we raided the target. Spy information always has to be challenged. If the spy tells you exactly where the target is and we get it, the spy is out of a meal ticket.

The facts you are given may not add up to reveal the whole picture, but only squares on a paint-by-numbers canvas.

During the 1991 Gulf War, President Bush's daily CIA briefer told the President that reports from General Norman Schwarzkopf were overestimating the numbers of Iraqi tanks and artillery being destroyed by our air attacks. CIA satellite photo analysts had come up with lower numbers. A huge bureaucratic battle broke out; Norm went ballistic; we set up a meeting in National Security Advisor Brent Scowcroft's office to sort things out; and I worked to pry Norm off the ceiling of his Riyadh headquarters.

The truth was, the CIA satellite photo analysts were not taking the whole picture of the battlefield into account; they were relying exclusively on narrow looking-down-a-soda-straw satellite images of the battlefield. Norm's assessment relied on several sources—expensive satellites, inexpensive pilot eyes debriefings, and low-level aerial photos.

A pair of experts from the CIA headquarters at Langley, Virginia, attended the White House meeting: a satellite photo expert and a multisource expert, who gathered his facts from a broad spectrum of sources, not from a single, narrow soda straw. His picture of the battlefield, in other words, was very like the one Norm was seeing and that other multisource analysts at CIA, the Defense Intelligence Agency, the National Security Agency, and the National Reconnaissance Office were seeing. I laid out my explanation of Norm's view from the field, and the multisource expert confirmed it: "Yes, that would be our assessment," he said. Norm's view prevailed.

Verified facts don't always come pure, but with qualifiers. My warning radar always goes on alert when qualifiers are attached to facts. Qualifiers like: My best judgment . . . I think . . . As best I can tell . . . Usually reliable sources say . . . For the most part . . . We've been told . . . and the like. I don't dismiss facts so qualified; but I'm cautious about taking them to the bank.

Don't get me wrong. I don't look down on intelligence gatherers, and I don't mean to condemn any specific intelligence staff or the intelligence community. It's a hard, stressful, vitally necessary job. During my career I've worked with intelligence agencies and experts of every kind, from a young lieutenant, battalion-level intelligence officer to all sixteen branches of the U.S. intelligence community. With rare exceptions, intelligence analysts do all they can to give you the information and facts you need to understand the enemy and the situation and come up with the best decision.

I found over the years that my intelligence staffs told the best story when I worked with them as they were putting it together. I questioned them constantly; I sent written analyses back, loaded with scribbles in the margins; I challenged them to defend their analyses. Staffs appreciated the challenge. They wanted to get the story right as much as I did.

Over time I developed for my intelligence staffs a set of four rules to ensure that we saw the process from the same perspective and to take off their shoulders some of the burden of accountability. The rules are simple; I'm told they hang in offices around the intelligence world:

- Tell me what you know.
- Tell me what you don't know.
- Then tell me what you think.
- Always distinguish which from which.

What you know means you are reasonably sure that your facts are corroborated. At best, you know where they came from, and you can confirm them with multiple sources. At times you will not have this level of assurance, but you're still pretty sure that your analysis is correct. It's okay to go with that if it's all you have;

but in every case, tell me why you are sure and your level of assurance.

During Desert Storm, our intelligence community was absolutely certain that the Iraqi army had chemical weapons. Not only had the Iraqi army used them in the past against their own citizens and against Iran, but there was good evidence of their continued existence. Based on this assessment, we equipped our troops with detection equipment and protective gear, and we trained them to fight in such an environment.

What you don't know is just as important. There is nothing worse than a leader believing he has accurate information when folks who know he doesn't don't tell him that he doesn't. I found myself in trouble on more than one occasion because people kept silent when they should have spoken up. My infamous speech at the UN in 2003 about Iraqi weapons of mass destruction (WMD) programs was not based on facts, though I thought it was.

The Iraqis were reported to have biological agent production facilities mounted in mobile vans. I highlighted the vans in my speech, having been assured that the information about their existence was multiple-sourced and solid. After the speech, the mobile van story fell apart—they didn't exist. A pair of facts then emerged that I should have known before I gave it. One, our intelligence people had never actually talked to the single source—nicknamed Curveball—for the information about the vans; he was a source whom some of our intelligence people considered flaky and unreliable. (They should have had *several* sources for their information.) Two, based on this and other information no one passed along to me, a number of senior analysts were unsure whether or not the vans existed, and they believed Curveball was unreliable. They had big don't-knows that they never passed on. Some of these same analysts later wrote books claiming they were shocked that I had relied on such deeply flawed evidence.

Yes, the evidence was deeply flawed. So why did no one stand up and speak out during the intense hours we worked on the speech? "We really don't know that! We can't trust that! You can't say that!" It takes courage to do that, especially if you are standing up to a view strongly held by superiors or to the generally prevailing view, or if you really don't want to acknowledge ignorance when your boss is demanding answers.

The leader can't be let off without blame in these situations. He too bears a burden. He has to relentlessly cross-examine the analysts until he is satisfied he's got what they know and has sanded them down until they've told him what they don't know. At the same time, the leader must realize that it takes courage for someone to stand up and say to him, "That's wrong." "You're wrong." Or: "We really don't know that." The leader should never shoot the messenger. Everybody is working together to find the right answer. If they're not, then you've got even more serious problems.

We need that kind of courage. We have to encourage it in our subordinates. I hate having to say, "Jeez, why didn't someone tell me?"

If I act on what you tell me you know and don't know, I am adding my experience and broader knowledge to yours. If my decision turns out badly, I am responsible, but so are you, and you should expect to be held accountable. Welcome to the real world!

In 1991, as we prepared for Operation Desert Storm, our intelligence people were sure the Iraqis had chemical weapons, but there were unresolved questions about whether or not they would use them. Some analysts and experts thought they would; others thought they wouldn't. It was a classic "don't know" situation. We thought they had them, and they had certainly used them. But they had to fear retaliation and worldwide condemnation, and it wasn't clear that their troops were still trained to use such weapons. I accepted this "don't know." We could have no certainty about whether or not they would use them until they used them . . . or tried to. And they had plenty of incentive not to.

Tell me what you think. Though verified facts are the golden nuggets of decision-making, unverified information, hunches, and even wild beliefs may sometimes prove to be just as important. Without wild beliefs there would be no stock market or hedge funds.

Your thoughts and opinions are vital, even if you can't prove or disprove them, and even if they are nothing more than hunches. You may be right. I have frequently found that someone's hunch is a more accurate view of reality than his knowledge. But if I act on your thinking or hunch, then I bear all the responsibility for the outcome, not you.

Many intelligence analysts and experts believed the Iraqis would use chemical weapons. That was their opinion. The facts could be taken either way. My own judgment was that they wouldn't use them. There was too much to lose. We had communicated to them that we would respond in an asymmetric way if they did, and we left them to imagine what that might be. They were aware of our capabilities.

I further believed that we could fight through any Iraqi chemical attacks. The possible effects back home worried me—public outrage and near-hysterical reactions. But I felt we could manage these. In making these judgments, I was relying on my experience and instincts. If I was wrong, the responsibility and accountability would be upon me and not the intelligence community.

It turned out that the Iraqis did not use chemical weapons.

Always distinguish which from which. I want as many inputs as time, staff, and circumstances allow. I weigh them all—corroborated facts, analysis, opinions, hunches, informed instinct—and come up with a course of action. There's no way I can do that unless you have carefully placed each of them—facts, opinions, analysis, hunches, instinct—in their proper boxes.

Years ago, one of my best friends, then Major General Butch Saint, got thrown out of the Army Chief of Staff's office for delivering bad news about one of the Chief's favorite programs. Butch knew before he walked in that he was entering the lion's den, and he wasn't surprised when he got thrown out. Word quickly spread around the Pentagon, as it always does when things like that happen. Not long after I heard about it I ran into Butch in a hallway. As we walked along, I offered him comforting words. "Hey," he said quietly, "he don't pay me to give him happy talk." I have never forgotten that. Butch retired as a four-star general.

Chapter Eighteen
Tell Me Early

There's an old Army story about a brand-new second lieutenant just out of airborne jumpmaster school who is supervising his first drop-zone exercise. He is standing there by the drop zone—a big, open field—watching the approaching planes. Standing next to him is a grizzled old sergeant who has been through this hundreds of times. The lead planes will be dropping artillery, trucks, and ammunition.

Everything is looking good and the lieutenant gives the okay to drop. The first chute comes out and deploys fully. The second one is a streamer and doesn't deploy. It hits the first one, which collapses. Subsequent chutes get caught up in the mess and they all start hitting the ground at full speed. Pieces of wreckage are flying everywhere, gasoline fires break out, touching off the ammunition and starting a brushfire that rapidly spreads into the surrounding woods.

The young lieutenant stands there contemplating the disaster. He finally says to the sergeant, "Umm, Sarge, do you think we should call someone?" His patient reply: "Well, Lieutenant, I don't rightly know how you are going to keep it a secret."

Staffs try like the devil to delay as long as possible passing bad news to the boss. That suits some bosses, but it never suited me. I had a standing rule for my staffs: "Let me know about a problem as soon as you know about it." Everyone knows the old adage: bad news, unlike wine, doesn't get better with time.

Knowing I had a problem was important, but it was more important to start the process of finding a solution. I always used that first notification to give the staff guidance about finding alternative solutions or about possibilities I didn't want them to consider. This was guidance, not a final decision. I always made it clear that I would not leap off a cliff or jump to a solution while they were still determining the shape of the problem.

But I still had to know about it. And I wanted all of it, not part of it. I wanted to hear all the bad. If they didn't tell me, I risked hearing it from some outsider or tripping over it myself. They knew they didn't want that to happen.

In 2003, American soldiers and interrogators in charge of Iraqi prisoners at the Abu Ghraib prison in Baghdad subjected prisoners to horrendous abuse, torture, and humiliation. Their actions were shocking and clearly illegal.

Late that year, one of the soldiers stationed at the prison reported the abuses to his superiors and said that photos had been taken by the abusers. The commanders in Iraq immediately took action and took steps to launch an investigation. Soon after that the news reached Secretary of Defense Donald Rumsfeld and General Richard Myers, Chairman of the Joint Chiefs of Staff, who told President Bush in early January 2004 that incidents at Abu Ghraib were being looked into. It seems that nobody told these senior leaders that these incidents were truly horrendous. General Ric Sanchez, the overall military commander in Iraq, announced the investigation on January 12. Soldiers were suspended from duty pending disciplinary action.

The machinery was working, but not all of it. The pipes leading up to the senior leaders were never turned on. The Abu Ghraib photos were available to senior Pentagon leaders, but it does not appear that Secretary Rumsfeld saw them, nor were they shown at the White House. A fuse was burning, but no one made the senior leadership aware that a bomb was about to go off.

In late April, CBS's *60 Minutes* broke the story wide open. They had obtained the photos and showed them on the air. The bomb went off and all hell broke loose.

I was shocked when I saw the photos. How could American soldiers do this? How could the implications of their eventually becoming public not set off alarm bells at the Pentagon and White House? Why was there no action at the top? Don Rumsfeld had been around a long time. If they had known what was going on, he and his staff would have immediately realized the dimensions of the crisis. So would the President's staff. And yet nearly four months went by and no one had elevated the material up the chain to the Secretary or the President.

If that had happened, the problem would not have been magically solved, but the people at the top would have had time to decide how to deal with the disaster and get to the bottom of it. The President was not told early.

Leaders should train their staffs that whenever the question reaches the surface of their mind—"Umm, you think we should call someone?"—the answer is almost always "Yes, and five minutes ago." And that's a pretty good rule for life, if you haven't yet set your woods on fire.

With early notification, we can all gang up on the problem from our different perspectives and not lose time.

As I have told my staff many times over the years, if you want to work for me, don't surprise me. And when you tell me, tell me everything.

Chapter Nineteen
Beware First Reports

DEWEY DEFEATS TRUMAN: probably the most famous incorrect first report in American history. The *Chicago Tribune* published this blazing, banner headline on election night, 1948, proclaiming that Governor Thomas E. Dewey of New York had beaten President Harry S. Truman and was now the President-elect. Wrong. A photo of President Truman holding up the *Tribune* and grinning like the Cheshire Cat vividly demonstrated the reality.

On the night of July 3, 1988, during my time as President Reagan's National Security Advisor, I got a call: the USS *Vincennes*, a guided missile cruiser, had just shot down an attacking Iranian F-14 fighter plane in the Persian Gulf. I called President Reagan to give him the report but cautioned him that it was a first report and didn't sound right to me. I couldn't understand why a lone F-14, primarily an air-to-air fighter, would dive on an Aegis cruiser on full alert, bristling with electronics and missiles designed to counter far more dangerous threats.

Not long after that, another report came in. The plane the *Vincennes* shot down wasn't an F-14; it was an Iranian Airbus passenger jet ascending on a normal flight path. The mistake sent 290 people to their deaths. The subsequent investigation concluded that the *Vincennes*'s commander should have been wary of the first report from his combat information center that a single combat plane was descending, when in fact an airliner was ascending. He should not have trusted the first report.

And then there were the March 2003 first reports claiming that CIA spies had located Saddam Hussein at Dora Farms. More first reports flowed in: "We know he is there, we know what room, we have eyes on it." It was worth taking the shot, and we attacked. Other first reports followed: "We see casualties coming out; we are sure we see Hussein being carried out." The reports that he was killed of course turned out to be wrong. I am not convinced that he was at Dora Farms that day. If he was, he escaped injury. But if the reports that he was there had turned out to be correct, the attack on Dora Farms would have been worthwhile.

In November 2003, the city of Tbilisi in the Republic of Georgia was in a state of rebellion. Flawed parliamentary elections had sparked massive street demonstrations demanding that the government of President Eduard Shevardnadze step down. Despite Shevardnadze's efforts to control matters, riots and violence threatened to break out. On the evening of the 22nd, Condi Rice, President Bush's National Security Advisor, called me with a report that the neighboring Russian Federation was sending in Spetsnaz special forces units to put down the demonstrators and restore order. A military intervention of this kind into the politics of a neighboring sovereign country would have just made things worse and might lead to civil war. I needed to check out this first report. It felt wrong to me. It seemed out of context.

The Russians had certainly demonstrated in the days of the Cold War that they would intervene militarily in satellite countries where there was danger that one of them might break free of Soviet domination. Hungary in 1956 was an example. Czechoslovakia in 1968 was another. But Russia and the world had changed radically since 1956 and '68. The Russians were following the situation in Georgia very closely, of course, but we had been in touch with them. I had received no indication that they were inclined to take military action.

I tasked my staff to get all the information they could and to check with the intelligence community for confirmation. I knew they would do their best, but I had to have this information fast. If Condi's report was true, we had a crisis; if it was not true we had to spike it as fast as possible before it got to the media and created a problem we didn't need with the Russians over why we had given credence to the rumor.

Meanwhile, my trusted colleague Igor Ivanov, the Russian foreign minister, was on his way to Tbilisi to try to mediate with Shevardnadze and opposition leaders. Igor had been one of Shevardnadze's deputies when the Georgian president had been the foreign minister of the Soviet Union. The two men were close, and I also knew Shevardnadze well.

Since Igor did not have a phone in his plane, I had to wait for him to arrive at the airport in Tbilisi. As soon as he landed he called on his cell phone. I lost no time telling him that we had heard the Spetsnaz might be moving in.

"Colin, that is nonsense," he replied. "Why would we do such a thing. I categorically deny it."

Igor was in the know, and I trusted him. I reported his denial to Dr. Rice; the first report quickly evaporated, and Igor continued his mediation efforts.

Later that day the Shevardnadze government resigned. New, fair elections soon followed.

We had to kill the first report quickly before it got off the runway, and that's what happened. The "Rose Revolution," as the November 2003 events in Georgia came to be called, succeeded. (The 2008 Russo-Georgian war demonstrated that the Russians are still capable of intervening militarily in the affairs of their neighbors.)

A first report follows every event. The first report may be entirely accurate and you can take it to the bank; it may be only partially accurate; or it may be totally wrong. How can we weigh first reports to determine where the best probability lies?

My experience with hundreds of first reports over the years has provided me with a mental checklist for reacting to them:

- Does it make common sense? Take a deep breath, rub your eyes.
- Does it fit in with everything else that is going on? Is there a context for this event?
- How much time do I have to figure this out?
- How can I confirm it? Launch the staff! Pick up the phone!
- What are the risks, costs, and opportunities lost if the report is true and we delay action?

- What are the risks, costs, and missed opportunities if it is false and we act too quickly?
- What are the stakes?
- Time's up! Do something! Keep searching!

I have experienced lots of first reports over the years that were true. Some I acted on, some I didn't. Some I regret that I didn't. You always have to remember that a clever enemy can create false first reports—some to influence you to action they want you to take, or some that seem so obviously false that you dismiss them when you should act on them. Hitler refused to believe first reports that the Allies would invade at Normandy.

In my own experience, a deep breath is always a good first reaction to a first report. Try to let a hot potato cool a bit before you pick it up.

Chapter Twenty
Five Audiences

Desert Storm in 1991 was the first cable news war. CNN flooded the airwaves with on-the-scene coverage; the broadcast networks followed; and satellite feeds broadcast the conflict to every corner of the earth. Hundreds of reporters showed up in the battle area wanting access to everything that was going on.

Those of us who were running the war had the additional task of satisfying the demand for news. Part of my job as Chairman of the Joint Chiefs of Staff, and Dick Cheney's job as Secretary of Defense, was to manage all this. In my view, we handled it well, despite considerable criticism.

I have always believed in the principle that the media's obligation to inform the American people imposed a duty on me to tell the media as much as possible about what was going on. I had the responsibility to help them understand our actions during Desert Storm so they could do their job. I also had an obligation not to give out information that could compromise our operations or put our troops at risk. Though the media invariably wanted to know more than I wanted them to know, and they criticized us for not telling them as much as they wanted to hear, they understood their job and mine.

We did a pretty good job finding the right balance between our two sometimes conflicting obligations. But we weren't the final judges of that. It was up to the American people to decide. They wanted all the news, but trusted us to protect their sons and daughters in combat.

Though this may not seem immediately obvious, *Saturday Night Live* provided a pretty good measure of how most Americans thought we were handling the press. A *SNL* skit portrayed Dick Cheney and me at a press conference where reporters asked us clearly over-the-top questions, such as "What time in the morning are you going to attack?" I think the people out there understood what we were up against and what we were trying to do.

Secretary Cheney and I gave a number of press conferences during the conflict. By then I had learned quite a bit about dealing with the press in our modern era of electronic news. Whenever I appeared before the press, I had come to realize that I was talking to multiple audiences and had to satisfy all of them. For most of my press appearances I identified five prominent audiences:

1. *The reporter asking the question.* The reporter is the least important audience. Always remember, you are not talking to the reporter, but through the reporter to the people out there watching and listening. That said, be respectful of the reporter. In an interview situation there is no such thing as a dumb question. Putting down a reporter makes you look like a bully.

2. *The American people who are watching and listening.* They want information, especially if their children and loved ones are engaged in the battle. They want to have conveyed to them a sense of confidence and assuredness that their leaders know what they're doing. They expect and deserve honesty. Included in this audience are our political and government leaders. Even if in they are in Washington, most of them are hearing this news for the first time.

3. *Political and military leaders in more than 190 foreign capitals.* Every one will have to comment and explain to their own people what you have said; in Desert Storm many of them had their own troops in the battle under U.S. command. That means that you are not just talking to foreign leaders, but to their fellow citizens and their families.

4. *The enemy, who is watching and listening carefully.* You don't want to give him anything he can use against you. You need to be an expert at sliding away from questions like "Is it true we don't have enough fuel to launch the operation?" . . . "Is it true that you are able to listen to Iraqi secure radio communications?" . . . "What about the report that you have special forces operating covertly west of Baghdad?" Some of our necessarily vague responses terribly disappointed reporters.

5. *Finally, the troops.* They have access to radio, television, print media, and now the Internet. You are talking about their lives. You never try to spin this audience. First, it won't work. Second, they are counting on you. They trust you, and you must never violate that trust.

If you are a senior leader—military, corporate, or financial—who plans to speak in public, you should make a thorough analysis of each audience you will be addressing. Be sure you are always talking through the questioner or the interviewer to the audiences who really matter.

I guess there are schools that teach these ideas, but my education came on the job. Sometimes we throw into the press breach a senior leader who has not yet completed his media on-the-job training.

I assigned Lieutenant General Cal Waller to be Norm Schwarzkopf's deputy during Desert Storm. I had known Cal for years; he was a brilliant officer; and I considered him one of my mentees. In December 1990, Dick Cheney and I arrived in Riyadh for briefings from Norm. We had with us a large press contingent pleading for information. Norm, Dick, and I were busy, so Cal, who had little press experience and had been in Riyadh for only a month, was tossed in to brief the press. Cal, doing his best to be forthcoming during the questioning, offered his view that we wouldn't be ready to attack until maybe mid-February. It was a big-time gaffe that contradicted what we and the President had been saying. And Cal had egg on his face. In fact the Air Force and Navy were already set to go by then; the Army needed more time. Meanwhile,

Cal's remarks became headline news around the world. The media couldn't believe their good fortune.

Cal felt terrible, but we reassured him; no damage was really done; and we were able to tamp down the uproar within twenty-four hours.

Since I was one of Cal's closest mentors, I wrote a note to him that night in my hotel about how to handle the press. It has applications far beyond a note between friends:

"Cal, with respect to the press, remember,

1. They get to pick the question. You get to pick the answer.

2. You don't have to answer any question you don't want to.

3. Never lie or dissemble, of course; but beware of being too candid or open.

4. Never answer hypothetical questions about the future.

5. Never reveal the private advice you have given your superiors.

6. Answers should be directed to the message you want readers/viewers to get. The interviewers are not your audience.

7. They're doing their job. You're doing yours. But you're the only one at risk.

8. Don't predict or speculate about future events.

9. Beware slang or one-liners unless you are consciously trying to produce a sound bite.

10. Don't wash dirty linen.

11. Do not answer any question containing a premise you disagree with.

12. Don't push yourself or be pushed into an answer you don't want to give.

13. If trapped, be vague and mumble.

14. Never cough or shift your feet.

15. When there are second follow-up questions, you're in trouble—break right, apply power, gain altitude, or eject."

As the years passed, I learned a couple of other lessons:

Thirty minutes is long enough for any interview. Any longer and you start to step on your own lines.

I never gave on-the-record interviews at a meal. You get too relaxed and think you are just hanging around with good friends.

Never shift in your chair, grab your ear, or touch your face. It's a signal that you have been caught.

Never pause to think of what to say. Start talking while you are thinking. You can always just repeat the question.

I learned the "you're the only one at risk" rule in 1987, when I was Deputy National Security Advisor, moments after my very first Sunday morning TV interview, on *This Week with David Brinkley.* I was doing fine; and we were near the end of the show. The great reporter Sam Donaldson, one of the regular panelists, grabbed the mike and in his aggressive manner asked, "Why should we trust you? You are a military officer, and after the recent NSC Iran-Contra scandal, with military officers in charge, why should we trust you?" During the half minute that was left, I thought I gave a rather good account of myself and why I could be trusted. After the show, I remarked to Sam that I thought I had gotten the best of that exchange. I won!

Sam smiled at my naïveté. "General," he said, "when you are with the press, you are the only one at risk. I can never lose." I never forgot that.

And he added, "Never smirk at us when you think you're ahead."

And never let them see you sweat.

PART V
Getting to 150 Percent

Chapter Twenty-One
What I Tell My New Aides

When I start out with a new front office staff, I have always found it useful to give them a sense of what I expect from them. They are nervous, anxious, wanting to please, but walking on eggshells. Here is a summary of my opening guidance, written down several years ago by a former aide but here somewhat modified. I have passed out the original list many times over the years:

HOW TO SURVIVE AS MY AIDE—
OR WHAT NOT TO DO

Don't ever hesitate to ask me what to do if uncertain. I trained my assistants never to act on instructions from me that they didn't fully understand. If you are confused, ask me to explain again exactly what I told you to do. If it still isn't clear, debate it with me to make sure you've got it. If you still aren't sure, then I am the one who is confused. I don't have a clear enough understanding of what I want you to do. It is time for me to sit and think my way through it again. Invariably, I find a fault in my analysis.

Don't ever sign my name, or for me.

I got this from an early boss of mine in the Pentagon, John Kester, a fastidious lawyer and a stickler for quality correspondence. Kester never let anyone sign his name or use an autopen machine. His signature created a legal document that had to stand up in court or anywhere else. Another habit of his: the only date he allowed on a document was the date it was signed. It was a legal document.

I followed those rules throughout my career. As Secretary of State it was my job to sign the elaborate commissioning certificates given to presidential appointees throughout the government who had been confirmed by the Senate. They came in by the dozens, but I signed each one, and when my son, Michael, was appointed chairman of the Federal Communications Commission, I added a smiley face to his.

I had to make some exceptions to this rule. Public mail during Desert Storm came in by the thousands. Since I couldn't sign every response, and believed every citizen who wrote me deserved and expected a response, I authorized one or two staff assistants to use the autopen machine for this purpose.

Never use money on my behalf.

I always gave my personal assistants a petty-cash fund to pay for daily incidentals. When they ran low, they asked for more. But never were they to go into their personal funds to pay for my lunch or for stamps or shaving cream. No exceptions. It is an abuse of position to allow otherwise. Never borrow from nor lend money to an assistant.

Avoid "The General Wants" syndrome— unless I really do.

"Jeez, you know the downstairs bathroom in my quarters is looking really shabby." Before you know what happened, someone has ordered a total renovation for $15,000 and a contract has been signed. A $20 gallon of paint was probably all that was needed. But now you will be answering to Congress about why you overspent on your government house. Worse, the *Washington Post* is working on a three-page exposé. Even inexpensive-seeming repairs get expensive when post engineers add to the cost a portion of their overhead fee for maintaining not only your house but the whole post.

Unless you are careful or protected by the people around you, a simple observation or aside can turn loose bureaucratic monsters wanting to please the general or the boss. How many CEOs have gotten nailed for a $75,000 conference room table? When I really want something, you'll know it.

Provide feedback, but be tactful to those who ask—talks between you and me are private and confidential.

Former New York mayor Ed Koch used to ask everybody in sight, "How'm I doing?"

Everyone wants feedback about how they are doing: what the boss thinks, how did the meeting go, is he okay or mad? Ed was asking his boss, the voters.

I need to know what my staff thinks about how I'm doing, and they need to know what I think about how they are doing. The information we exchange is private and privileged. What goes on in the office stays in the office.

Every organization loves, thrives on, and solicits gossip. And subordinates need candid feedback on what the boss thinks and feels, without asking the boss. Every effective leader communicates his feelings and reactions to the organization. But there are times when the criticism, praise, or course correction should not come from the leader but from someone close to him and seemingly empowered to speak for him. After one of my subordinates presented me a particularly poor briefing, my assistant might be asked, "Hey, how mad was the General?"

"Not to worry, he knows you'll fix it by next week."

Or, "I've never seen him so pissed. Man, you had better get your act together by the end of the week or he's gonna drop you like a rock."

The trusted assistant may have to occasionally fib or sweet-talk for the good of the organization. Egos are tricky things and must be carefully managed and massaged.

I'm no exception. I need to know how I'm doing.

I live on the speaking circuit. After each speech I never ask my client how I did. I have my assistant at the speaker's bureau ask the client's assistant how the event went and how did the General do. Assistants love to gossip and I get useful feedback. When we don't get an answer, I know I was not at my best and had better review what I told them.

Alma and my family have nothing to do with the office. Never interrupt with calls from Alma unless there's a crisis.

I love my family deeply and passionately. They are my life . . . but not all my life. Their place in my life does not extend to the office.

My wife, Alma, is aware that good fences make good marriages. She runs the house, me, and the kids. I run the office. She never gets involved in policy, personalities, gossip, or anything else at the office. She doesn't cross-examine me over dinner about what's going on. We had been married ten years before she could tell one rank insignia from another. She never forgot what General Bernie Rogers, Army Chief of Staff, told us when I went through charm school before being promoted to brigadier general: "Over the next couple of years, I'll have to bounce several of you out because your wives will start acting like they are generals. I know you don't believe me now, but just watch." He was right. Alma is smart as the devil and always knows what's going on, but we always preserved that strict boundary at the threshold of our home. She knows when she calls the office that I may not take the call right away.

What's true of Alma is true of the rest of my family.

Never keep anybody waiting on the phone—call back.

I came up with this one before the days of voice mail, call waiting, call forwarding, and that modern nightmare, phone trees. I just thought it was rude to put someone on hold and leave them for more than a moment. You are wasting their time. If you can't put them through right away, tell them we'll call back. And then make sure we do. I tried to never end the day without answering all calls.

When I became Secretary of State I kept my doors open so I could keep track of what was going on in the outer offices of the suite. Too often I could hear the phones ringing until they got shunted into voice mail. Unacceptable. The front office of the Secretary of State cannot use voice mail. The phone will be answered after the third ring by a welcoming human, even if that human is me. And you really don't want me to have to answer the phone. I want every call dealt with or directed for real to some place in the department where it can be dealt with. I want people to say, "Wow, I called the Secretary of State's office and I got right through."

I like meetings generally uninterrupted. I ask a lot of questions. I like questions and debates.

If I am having a meeting, I am having a meeting. Meetings should be open and probing, and they should be sacred time.

I like meetings that dive deeply into the issue at hand. You can only do that if everyone there feels free to ask questions that peel away the skin and get down to the core of an issue. And I like to be challenged.

Don't assume that I already know the answers. If I did, I wouldn't need a meeting.

Meetings are sacrosanct. Don't break in to tell me someone is calling or someone needs me. I believe in respecting the others at the meeting. I don't want to waste their time. Their time is as valuable as mine. If you interrupt, it had better be urgent.

I'm a people/phone junkie. I like to remain enormously accessible.

The higher I rose, the harder I had to work to remain accessible to people strolling by and not be walled off by staffs and doors.

At State, I had wide-open doors. I worked in a small inner office with a beautiful formal meeting area between me and the office staff. I could keep track of what was going on, listen for the phone, and pick up giggles and snickers I could ask about later; people who needed to see me could look in to check if I was busy. Even when I have to keep my door closed, if someone needs to see me, just give me a heads-up and if I'm free, send them in.

I will develop ways of getting to know what's happening.

So, don't hesitate to tell me. The more senior you become, the more staff you have to protect you from yourself and to push their own agenda. They mean well, but they can insulate you from ground truth. You have to get out and walk the floor. Have trusted agents and friends call you when they think the emperor has no clothes. In the Army, chaplains, inspectors general, and sergeants major can always give you a ground truth perspective. Above all, never forget you were ground truth once. Never lose that bond with what is happening down in the subbasement.

Don't accept speaking engagements without my knowledge.

In fact, don't accept any calendar commitment without my knowledge. I am a nut about my calendar. I must control my time. It's the only real asset I have. No appointment is accepted, no event is scheduled, without my personal approval. But unscheduled time is fair game. I try to keep my door open. If you really need to see me and I'm not busy or taking a nap, come on in.

Keep accurate calendars and records. And keep faithful track of calls and whom I have seen. I'll always return calls.

The older I get the more important this one has become. My memory remains near perfect; it just operates a lot slower than it once did. Accurate calendars and phone logs have proven invaluable. They have rescued me on many occasions.

On the other hand, whether or not to keep a diary is not an easy question to answer these days—especially for people with a significant public presence. The information revolution, emails, the Freedom of Information Act, and now WikiLeaks suggest caution in keeping personal records. Such records are important for historical purposes, but I fear a lot is being lost to historians because of the sensible exercise of caution and discretion.

I once found myself in a federal court as a witness, brought in to try to clarify a few ambiguous words I had jotted down in a calendar book. Because they were just a few words and not a narrative, the lawyers on both sides had fun deposing me and filling in the blanks to their advantage.

I tend to get moody or preoccupied. I will snap, but that clears the air.

I try to sail on an even keel, but I'm human. Sometimes I am so consumed with an issue that I am oblivious, and even rude, to people around me. Sometimes I let my temper get the best of me, and I blow up. Leave me alone, stay out of the way, and I'll be back to normal shortly. Don't take it personally; I really just got hung up on the issue. I can't have people around me who go fetal when they get caught on my gun-target line.

Be punctual; don't waste my time.

Punctuality is a sign of seriousness, discipline, understanding the value of time, and simple respect and courtesy. One of the first things you learn in the Army is to be on time. In the Army being on time is no casual thing. You'll remember the old war movies where leaders synchronize their watches so everyone can jump off at the exactly correct time. I was taught at the Infantry School to never be without a watch, a pen, and a notepad.

I insist on punctuality. The meeting starts when I said it would, with or without you. Show up late for an appointment and you might find it canceled. Meetings that start late and go too long waste everyone's time.

Emergencies can arise, and meetings can start late or go long; but you can't be late just because you didn't organize your day. When people are kept waiting just out of indiscipline or poor time management my blood boils.

The media wrote stories about President George W. Bush closing the door on you if you weren't at a meeting on time. One day I was delayed by an important phone call and was about a minute late to a cabinet meeting. The door was locked. It was then opened and I slipped in. Everyone laughed, but it was also a lesson. Some reporters on a slow day started a story that President Bush was dissing me. Nope, he did it all the time to others . . . even to Karl Rove, to my delight.

I prefer written information to oral. Writing encourages discipline.

I *do* love oral arguments and presentations. Sharp, well-organized briefings are good. But writing trumps oral. A well-written analysis of an issue, listing the alternatives and the opposing points of view, distributed and studied in advance, makes for a more productive meeting. In the quiet solitude of my home or office, I can find inconsistencies and weaknesses or strengths. I am then ready to hear the oral arguments, sufficiently armed not to be influenced solely by the performance skill of the protagonists.

I do lots of paperwork—and I like doing it.

I was always a good staff officer. I would normally grind through dozens of papers a day. I read fast. I know how to scan and how to weed out filler. And I enjoy it. I made it a habit in all my senior jobs to respond to all papers the same day they were sent to me. On those occasions when I couldn't get to something the day it reached me, I'd work on it at home, and respond the next day. If you didn't get it back within twenty-four hours, it wasn't because I didn't get to it. You better worry about why you didn't. It usually wasn't good news.

Make sure correspondence is excellent. No split infinitives.

I want the best possible correspondence. Human writing. Avoid stilted, puffed-up, hyperbolic, over-adjectived bureaucratic claptrap. Write to me as you would talk to me. Another lesson John Kester drilled into me was a hatred of split infinitives. Split infinitives have become more acceptable in recent years (and my writing is not free of them), but I still insist on eliminating them to force my folks to read every line carefully. Getting rid of split infinitives is not what counts. What counts is reading every line carefully.

Never, never permit illegal or stupid actions.

Any questions?

No surprises. I don't like to be blindsided. Bad news doesn't get any better with time. If there is a problem brewing, I want to know of it early—heads-up as soon as possible.

If you run on an even keel, and have a good team whom you trust and who trust you, then your folks won't be afraid to share the worst with you as soon as possible. Your team will know they can bring you a problem without you blowing up and that you'll give them guidance to start working the problem. Bonds of mutual trust and confidence among all the players will take you a long way down the road to a solution.

Speak precisely. I often fudge for a reason. Don't overinterpret what I say.

I like short declarative sentences with lots of protein and no fat. I try to speak precisely. Don't read more into what I say than what's there. And don't be upset if I fudge. The fudge is part of the precision. Don't try to interpret it, expand it, or contract it. I have always fudged for a reason, even if you don't know what it is.

Don't rush into decisions—make them timely and correct.

Time management is an essential feature of decision-making. One of the first questions a commander considers when faced with a mission on the battlefield is "How much time do I have before I execute?" Take a third of that time to analyze and decide. Leave two-thirds of the time for subordinates to do their analysis and make their plans. Use all the time you have.

Don't make a snap decision. Think about it, do your analysis, let your staff do their analysis. Gather all the information you can. When you enter the range of 40 to 70 percent of all available information, think about making your decision. Above all, never wait too long, never run out of time.

In the Army we had an expression, OBE—overtaken by events. In bureaucratic terms being OBE is a felonious offense. You blew it. If you took too much time to study the issue, to staff it, or to think about it, you became OBE. The issue has moved on or an autopilot decision has been made. No one cares what you think anymore—the train has left the station.

I have found over the years that my new staff welcomed these rules. They got us all playing from the same sheet of music. But they are not my only technique for getting a new team in harmony with each other and with me. This one almost always upsets them, but the indigestion goes away quickly:

"In our first weeks together," I warn them, "I will drive you to drink or worse with my constant corrections and nitpicking. You will think you are not doing well. But in a few weeks, as we adjust to each other, the little notes stop, the corrections are fewer, and we will settle into a comfortable pattern."

I use my torture technique to speed up the process.

Your staff needs to become your family as quickly as possible. Create a constructive environment and you'll have a winning team that will not let you down. In all my years as a general or senior official, I never had an IG complaint filed against me.

Chapter Twenty-Two
One Team, One Fight

When General George Joulwan was our Southern Command commander some years ago, he ended all of his messages with the slogan "One Team, One Fight." He greeted you in person the same way. After a while we started smiling whenever we heard George's slogan. But it was a good idea—worth taking to heart. It was a constant reminder to his command that everyone had to come together as a team to prosecute a fight that everyone agreed had to be won. It remains a good idea.

I tried to capture that spirit as Chairman of the Joint Chiefs of Staff. The Chairman is an advisor and commands nothing. He works through influence and persuasion. The other members of the Joint Chiefs of Staff are also advisors, but the Chiefs of the Army, Navy,

Marine Corps, and Air Force also have large organizations to run and protect. It was important for me to understand this duality of responsibility, recognizing not only their role as service chiefs, but also their larger duty as members of the Joint Chiefs.

I worked hard to create a sense of "One Team, One Fight." I commissioned a manual to capture this spirit. In its preface I wrote the following:

> When a team takes to the field, individual specialists come together to achieve a team win. All players try to do their very best because every other player, the team, and the home town are counting on them to win.
>
> So it is when the Armed Forces of the United States go to war. We must win every time.
>
> Every soldier must take the battlefield believing his or her unit is the best in the world.
>
> Every pilot must take off believing there is no one better in the sky.
>
> Every sailor standing watch must believe there is no better ship at sea.
>
> Every Marine must hit the beach believing that there are no better infantrymen in the world.
>
> But they all must also believe that they are part of a team, a joint team that fights together to win.

This is our history, this is our tradition, this is our future.

Fast-forward a few years to the State Department. The State Department consists of Foreign Service officers and the specialists who support them—civil servants and the Foreign Service local nationals who support our embassies. The Foreign Service officers are the most widely known. They are our diplomats and ambassadors—elite experts. Our Civil Service consists of professional and enormously capable support personnel.

Every year we observed Foreign Service Day, a day when retired Foreign Service officers returned to the State Department for rebonding and briefings.

I wanted to penetrate the cultural and other boundaries that existed between Foreign Service and Civil Service employees. With that in mind we introduced leadership training for mid-level and senior Civil Service managers and took other steps to emphasize their importance.

As part of that effort, I decided to change Foreign Service Day to Foreign Affairs (FA) Day and to invite retired Civil Servants to attend.

Whoops. We got noise from the Foreign Service community. They felt something was being taken away from them. There were mutterings that many would not attend. We worried about the turnout, but on FA Day, the auditorium was filled with Foreign Service officers and a significant number of Civil Servants. No one's ox got gored. And the Foreign Service realized the value of this kind of bonding. "One Team, One Fight."

Every good leader I have known understands instinctively the need to communicate to followers a common purpose, a purpose that comes down from the leader and is internalized by the entire team. Armed with a common purpose, an organization's various parts will strive to achieve that purpose and will not go riding off in every direction.

I have also seen many organizations that resemble nothing less than warring tribes. They usually fail.

Chapter Twenty-Three
Compete to Win

The military encourages competition. War is a competition, the ultimate test of purpose, preparation, determination, courage, risk, and execution. Business is a competition. In fact, in almost every human endeavor where there are two teams, groups, or sides, there is a competition.

People need to test themselves, prove themselves, not just to show that they are better than the other guy or the other team, but to show that they have trained and raised their skills as high as they can. Winning is great, and always better than losing, but perfecting our skills and capabilities is great, too.

In 1986, when I commanded the V Corps in Germany, the corps participated in two major international military competitions. One was called the

Boeselager competition; the other was the Canadian Cup. These were World Series events. Boeselager was an annual competition to pick the best NATO cavalry troop. The Canadian Cup was an intense competition to determine the very best tank platoon. You weren't competitive in these tests without putting forward an extraordinary effort to prepare the crews. Once you designated the unit that would represent you, every effort was made to find the corps' best leaders and experts and transfer them to those units. You then gave them priority for training ammunition, access to firing ranges, and whatever other resources they needed. Other units had to sacrifice for these Super Bowl–level competitors.

A case can be made that this kind of competition is not healthy. You don't go to war with your Super Bowl team but with every team in the league. Nevertheless, I did whatever it took to win, within the rules. I didn't like the idea of shorting my other units, but once you decide to go for a win, you give it all you have. You mass your resources, you explain to those being shorted why that must be done, and you go for the win.

Although I was transferred to the White House before the competitions, the teams we put together went on to win both events. No one corps had ever won both in the same year.

A more down-to-earth example occurred earlier in my career, when I was a battalion commander in Korea.

Every day, I set aside time to walk through the battalion area checking things out. One day, I saw one of my soldiers approaching from the direction of brigade headquarters. He looked a little down and was wearing his dress uniform rather than our normal fatigues. He saluted, and I asked him what was wrong, fearing he was just coming from a court-martial.

"I've just been in the Soldier of the Month competition," he told me, "before a board of senior sergeants."

"How'd you do?" I asked him.

"Not good. Sorry, sir."

"Thanks for your good try, soldier," I told him. "Too bad it didn't work out." I felt a lot of sympathy for him. "By the way," I asked, "when did you learn you were going before the board?"

"Last night."

I patted him on the back and went straight to my office for a come-to-Jesus session with my command sergeant major and first sergeants. "We will never do this again," I told them. "We will never throw our soldiers into a competition or into a battle, any battle, without preparing them and taking the necessary time to get them ready to win. That's what leaders do; we prepare our troops."

Our battalion won the Soldier of the Month competition for the next few months, until the other battalion commanders caught on and put in their own best effort.

Soldiers given a task they haven't been prepared for lose confidence in themselves and, fatally, in their leaders.

But sometimes you can be surprised.

In 1976, I commanded the 2nd Brigade of the 101st Airborne Division. We put together a team to compete in the division's annual boxing competition—a pretty good team and a heck of a coach. We set out to win, but we had one missing link. We didn't have a boxer who could compete in the featherweight class (120–125 pounds). Not until my adjutant, Jim Hallums, came in one day. He'd found a very small young soldier, Pee Wee Preston. Pee Wee had never boxed, and he was tiny. He would qualify for the featherweight competition. The real hook was that no other unit had a soldier who could make the weight. We would win the class by forfeit. We asked Pee Wee if he was willing to be on the team; he would probably never have to fight. He agreed to do it for the brigade . . . especially after we assured him that if he did, he would not have to go to Panama with his battalion for jungle training. Pee Wee was deathly afraid of snakes.

We insisted that he train just as hard as everyone else. He was taught the basics of boxing; he hit the bag, sparred, jumped rope, and did everything everyone else did.

The week of the tournament arrived, and our team was doing well. Pee Wee got in the ring twice, got the forfeits, and we got the points. But on the third night, disaster struck. One of the other units found, or imported, a Panamanian featherweight who was a miniature near double for the great Panamanian boxer Roberto Duran. This guy was going to fight Pee Wee. Yikes.

We told Pee Wee he could forget the deal; he didn't have to fight. But he wanted to go ahead. His whole battalion was leaving for Panama late that night, and they were in the stands watching. He couldn't let his guys down.

Pee Wee got in the ring, and the other kid raced across and proceeded to whomp up on him. Pee Wee never threw a punch back, but he took the other kid's punches, keeping his arms up, protecting his head and body, the way he had been taught, and he made it through round 1. Our side was cheering tentatively: "Attaboy, Pee Wee! Hang in there, kid!" Round 2 was a repeat, but he kept going. He was in shape, and he wasn't getting hurt. The other guy was looking winded and frustrated, just from the sheer effort of pounding on Pee Wee. The cheering for Pee Wee had grown a lot louder. He hadn't thrown a punch, but he was game. He had spirit. Round 3 opened and the other kid came out slowly. He was tired and weakened from beating up on Pee Wee. He was not in shape! You know what comes next: Pee Wee landed a single punch, and the other guy dropped his arms and quit—a TKO for Pee Wee.

His buddies went nuts in the stands. Pee Wee was the 101st Airborne Division featherweight champion. He had been prepared for a fight we never thought he'd have to fight. But he had been prepared enough to win.

Later, unfortunately, when we went to Fort Bragg, North Carolina, for the XVIII Airborne Corps competition, the 82nd Airborne Division entered a real boxer, and Pee Wee lost. But no matter—he represented himself and us well.

There are many kinds of competition. You can have a constructive competition that goes beyond just finding a champion. I am a believer in lots of intramural competitions within units. Best supply room, best soldier, best clerk, best armorer, you name it. Do it every month, and do it with standards that make it possible for anyone putting forth the effort to win.

Without competition, we all become dull, unfocused, and flabby—mentally and physically.

Chapter Twenty-Four
Swagger Sticks

"That's an order!" has long been a cliché movie line, usually blasted out forcefully by duff, blustering generals, commonly referred to as "the Brass." I hated the term, as much as I came to hate the term "striped-pants diplomats" when I became Secretary of State. Stereotypical images are hard to bury.

In my thirty-five years of service, I don't ever recall telling anyone, "That's an order." And now that I think of it, I don't think I ever heard anyone else say it. Yes, there are times when you want your instructions carried out without further discussion and carried out immediately, despite any reservations or reluctance. Just tell them to do it.

But there are often better ways to get what you want done than to huff and puff and bellow out an order.

The leader must impose his will. Clever, gifted leaders, in sync with their units and culture, can often command with the most delicate touch. Time permitting, it is far better to gain buy-in from followers by explaining what you are trying to achieve and the important role they are about to play in accomplishing the mission. The American soldier is better led than driven.

General David Shoup was Commandant of the Marine Corps in the early 1960s. Although the things were an anachronism, it was still common back in those days for officers to carry swagger sticks or riding crops, a custom left over from our British colonial heritage. You see British officers carrying swagger sticks in World War II movies, and you might still see the tradition practiced in Commonwealth nations. I had a swagger stick back when I was a young lieutenant. I treasured it. Sergeant Artis Westberry, my instructor in ROTC summer camp in 1957, made it for me, and I proudly carried it to point out things to soldiers and to beat the side of my leg.

Even way back in those days, swagger sticks were slowly going out of fashion in the Army, but the Marines persistently held on to the tradition. General Shoup thought it was time to get rid of them. As Commandant, he could have just put out a one-sentence order banning the silly things. But Shoup was a very wise leader and took a slightly different tack. He put out an instruction that simply said: "Officers are authorized to carry swagger sticks if they feel the need."

The sticks were gone overnight. I often wonder if he was laughing when he came up with that sentence. He knew his Marines. "We don't need no stinkin' swagger sticks."

Every organization has "swagger sticks" that are deeply rooted in its culture. Yes, you can just wipe them out, but it is usually not hard to find a way to expose them as anachronisms and put them out of their obsolescent misery, to the delight and support of all.

Chapter Twenty-Five
They'll Bitch About the Brand

Many years ago when I was a junior officer, we were looking for ways to improve morale, get in tune with a new generation of young soldiers, and cut back the number of soldiers who were drinking too much and getting arrested for DUI or, worse, getting into accidents.

Somebody came up with the idea of installing beer machines in the barracks so troops could drink, if they chose, right at home. Our sergeants didn't think this was a great idea. Unrestricted access to beer would encourage unrestrained drinking and result in rowdy behavior and beer brawls in the barracks.

The troops thought it was a great idea, predictably, and they pressed for it. No decision came . . . which set loose lots of bitching.

Would installing beer machines end the bitching and improve morale? Many of us thought so.

One of my savviest sergeants quietly pointed out to me the flaw in that thinking. "Lieutenant, putting machines in the barracks won't end the bitching. They'll just start to bitch about the brand of beer in the machines, except they will be drunk when they bitch."

We didn't put in the machines. And today's Army has worked hard to keep alcohol away from troops. It's a better, safer Army.

The big lesson I learned from this little episode: as you examine solutions, make sure you think them through down several levels into secondary effects, and when you arrive at what you believe will be a solution, you have to then ask yourself if you have the real solution, or if you have just let wishful thinking set you up for more problems.

This lesson applies to all kinds of problems, large and small. And bitching about brands can take place in all kinds of circumstances. Sometimes these are deadly serious. Let's change the scene from beer in barracks many years ago to the invasion of Iraq in 2003.

In 2003, we marched up to Baghdad, the city fell in days, and the regime of Saddam Hussein collapsed. We saw these victories as a great success and the end of a big problem . . . with little thought given to what we would have to take care of once we had achieved victory.

Would opening the door to freedom bring stability and peace to that tragic country? Many American leaders thought so.

Too bad they didn't have some savvy sergeant to quietly point out that we hadn't answered the question about how the changes we started would affect the people of Iraq or the makeup of Iraqi society, which, it turned out, is a jumble of sectarian brands. Iraqis have been bitching about these brands for centuries. Their new freedoms didn't stop the bitching, sparking disagreements and conflicts that turned our wonderful instant success into a terrible, nagging crisis. It took us years to achieve enough stability for American troops to be disengaged. For years wishful thinking drove a flawed strategy. Meanwhile, the argument in Iraq over brands continues and is liable to do so for years to come.

I learned a second lesson from the beer in barracks episode: surround yourself with sergeants—that is, people with ground truth experience whose thinking is not contaminated with grand theories.

Before we invaded Iraq, we should have listened to more people with ground truth experience in the region (these people were out there) and fewer idea-heavy, big egos in Washington.

Chapter Twenty-Six
After Thirty Days, You Own the Sheets

In the old days before computerized and centralized management systems, taking command of a rifle company was a far more interesting and personal process than it has since become. All the property in the company was registered in a company property book, an ordinary ledger with entries written in ink. Before assuming command of the company, the new commander and the outgoing commander would conduct an inventory of all the property. Every rifle, bunk, chair, desk, sheet, and pillow had to be accounted for. If anything was missing, the outgoing commander had to search for it and find it, pay for it, or seek relief through a process known as a "Report of Survey." After signing for the property and taking command, the new commander had thirty days to discover anything else that was missing or any other discrepancies. If during the thirty-day window he found anything missing, the new commander could initiate action that would either lay the problem on his predecessor or relieve the new commander of accountability. (Once I came up short on sheets and found what I needed at the post mortuary.)

As we used to say, "After thirty days, you own the sheets." On day thirty-one any discrepancies or shortages became your problem.

I loved this stark, clear way of assigning responsibility and accountability. No whining, no complaining, and no blaming the guy you replaced. Above all, don't waste time trying to cop a plea or blame the other guy. Too late, you've had your grace period. You own the sheets.

At levels above a small rifle company, there is a more sophisticated (tongue in cheek) way to handle these transitions. It's called the "Three Envelopes Construct." The outgoing leader gives the new leader three envelopes—labeled "Envelope 1," "Envelope 2," and "Envelope 3"—and tells him to open them in order if he runs into trouble. The new leader launches in a blaze of glory. But after a month or so, troubles start landing on him. He opens the first envelope, and the note inside says: "Blame me." So he goes around complaining about the mess he inherited. Things settle down, but a couple of months later he is back in trouble. He opens the second envelope: "Reorganize." He immediately starts a major study to determine the kind of reorganization that would improve the situation. For months, the reorganization study moves all the boxes and people around and creates a new paradigm. Everyone is distracted. The new paradigm looks exciting, but nothing is solved and everyone is confused.

The now no longer new commander is in dire straits and beside himself with worry. In desperation he opens the third envelope. The note says: "Prepare three envelopes."

The Three Envelopes Construct does not work with elected politicians. They will blame their predecessors as long as they can. If things are going wrong, it is not their fault. If things are going well, it is only through their superb efforts to fix the mess they inherited. If their predecessor comes from their own party they may have to complain sotto voce.

For normal mammalian human beings in a line position, assume your predecessor did a good job, and if he didn't, be silent. Move onward and upward. You are in charge. Take charge. And always remember, "You now own the sheets."

Chapter Twenty-Seven
Mirror, Mirror, on the Wall

I am pretty good at knowing and analyzing my strengths and weaknesses; but I keep the latter private. Though I never share these with anyone, my family and friends are quite willing to tell me what they are in detail. Self-examination is tough and worse when your friends and family join in. I am so glad that 360-degree evaluations came into vogue long after I stopped being evaluated. During the process, your ego is vulnerable, your self-respect challenged, your decisions questioned, and your fallibility made manifest. Still, such examination is essential to improving yourself, getting in better touch with the people in your life, facing your demons, and moving on. Looking deeply into a mirror and seeing an accurate reflection is therapeutic and healthy.

If it is difficult for individuals, it is even more difficult for groups of individuals in an organization with superiors and subordinates, where candor can put members of the group at risk, or where your honesty may be seen as disloyal or get you condemned as failing to be a team member. An organization that is unable to create the environment for this kind of evaluation is an organization that is holding itself back. The challenge is to get beyond the personalities, the egos, and the tendency to be blind to unpleasant conditions and move forward with no feelings bruised beyond repair. This is a real test of leadership and confidence in the team and the bonds that hold a team together.

Honest, brutal self-examination is especially difficult, but even more vital after a mess, a screw-up, or a failing performance. The Army faced such a crisis after the Vietnam War. There were no victory parades, and the nation got rid of the draft and distanced itself from the nascent all-volunteer force it had launched. We were in the midst of the countercultural revolution, racial and drug problems, and a shaken political system that would see the resignations in disgrace of a president and vice president. We had to reform an institution with deep cultural roots and a proud history—an institution that had recently failed to achieve its ultimate reason for existence, success in war. We set about rewriting our doctrine, reorganizing our units, and training all-volunteer recruits, many of whom were deficient in education or had behavioral issues. For me, it was the most demanding, exciting, and rewarding time in my career. We succeeded and rebuilt a first-class Army, as good as any that went before.

One of the most powerful tools the Army used to achieve this success was a technique called the After-Action Review (AAR). The AAR concept was first tested and proved at the newly established National Training Center (NTC), at Fort Irwin, California, arguably the most innovative training facility ever created. Both the NTC and AARs are still going strong.

The NTC consists of 600,000 acres of rolling desert, ideally suited for mechanized maneuver training and live firing in an utterly realistic environment. Units coming to the NTC to train face off against a highly skilled and trained enemy—called an Opposing Force (OpFor)—that is stationed at Fort Irwin. Both the good guys and the bad guys are wired, so their actions can be followed on computers at a centralized control center.

Training against a simulated enemy is not new. Armies have been trying to make training realistically close to actual combat for a long, long time. What makes the NTC unique is the comprehensive AAR that follows the completion of every battle. At AARs leaders, observers, and evaluators sit in the control center and watch the battle replayed like a video game. Every vehicle moving across the battlefield can be identified; every movement of troops and vehicles, every action, every gunshot has been recorded and can be replayed in several ways. For instance, the actual battle can be superimposed over the commanders' plans, comparing and contrasting reality with expectations. I have watched many an aspiring Patton put his original plan up on the screen and then watch his tanks and armored vehicles go wandering off in the wrong direction, firing at each other, as the OpFor rolls up his flank and defeats him. It reminds the young Patton of two military maxims: "No plan survives first contact with an enemy" and "Even the most brilliant of strategists must occasionally take into account the presence of an enemy."

All of this is then exhaustively analyzed. Nothing is held back, nothing is ignored. During the review, leaders, observers, and evaluators come together to present their own assessments of how they saw the battle unfold and why they made their decisions and took their actions.

The purpose of the review is to autopsy the exercise, not to give a grade or to anoint the commander as a future Patton . . . or Custer. Learning and improvement are the sole focus, not the unit's success or failure in the mission. It's not a blame game.

After the review, the subordinate leaders are then expected to go back to their units and share the AAR results down to the last soldier. Each subordinate unit conducts its own AAR.

The AAR system works because it is a training process, not an evaluation process. That doesn't mean feelings won't be hurt or unfavorable impressions created. The needs of the mission must come first. Though AARs are not about assigning blame, poor performance over time will naturally be noticed. A commander who consistently does poorly, or worse, is probably not suited for the job, or for command at a higher level. Those who consistently do well get noticed.

Because it works so well, the AAR system has been extended to all training throughout the Army. Watching AARs, I witnessed the birth of a new Army focused less on proving your worth by scoring points than on training our soldiers to be more effective. In my early days in the Army, evaluations were generally a matter of mechanically working through stylized teach-to-the-test checklists. Today, the system asks, "Where do we need more training? How do we make our troops better and more skilled?"

The result of the new training system was demonstrated in Operation Desert Storm in 1991 and subsequent conflicts. After engaging in actual combat, officers and soldiers who had been through the NTC experience reported that it had replicated down to nitty-gritty details the demands of real combat. It gave them a decisive edge when they faced the Iraqi army.

NFL teams go through a similar process after each week's games. They review their own game films, they review films of their next opponent, and they constantly ask themselves, "What can we do to fix our mistakes and improve?"

The AAR process is applicable to any organization that truly wants to know how it is doing, where it needs to improve, and how it can get to the bottom of a problem or dispute. What have we done right? What have we done wrong? The sole goal is to improve our performance. It's not about your ego or mine. If we are a team, we can level with each other in a spirit of "how do we do better?" We will not cover up mistakes, reorganize around them, or stare at the sky. It requires honest participation, a focus on learning, and a commitment not to use AARs as a means to assign grades. High-performing organizations understand the need for this kind of evaluation. I have also seen others whose leadership doesn't have the guts to look into the mirror. All of us have seen in recent years too many pitiful examples of companies and organizations that live and succeed in the moment and refuse to see the reality of the fuse burning in the basement. Leaders should never bury a problem; you can be sure it will eventually rise from its grave and walk the earth again.

I have tried to apply the AAR philosophy in all my post-Army assignments. During my days as Secretary of State, I was responsible for submitting an annual report to Congress about trends in terrorist incidents. The report was prepared by the CIA, reviewed by my staff, and sent to Congress in my name.

One year, Congressman Henry Waxman of California attacked the report. He accused me of understating the terrorist problem and of cooking the books by reporting fewer worldwide terrorist incidents than he believed the data showed. My staff initially circled the wagons and defended our position—the traditional bureaucratic response. But I wanted to find out who was right. If Congressman Waxman was right, we had to make changes, and do that before we had to defend our position before an open congressional committee. If we were right, I was ready to take on Henry, a good friend as long as we weren't across the table from each other at a televised hearing.

At my staff meeting the next morning we conducted an AAR. I wasn't happy with what I heard. Rather than starting at the beginning and analyzing exactly how the original report was generated, the staff just tossed up justifications for the report we had printed and distributed.

I told them to look at Congressman Waxman as though he were our OpFor at the NTC; his negative evaluation of our report was equivalent to an OpFor victory in an early engagement. I thought we should listen to his criticism, concede that he might be right, and fix the problems he'd spotlighted so they would not end up in lurid display before his congressional committee. That would have been equivalent to losing the final battle.

At another AAR the next morning, we brought in everyone involved in preparing the report, and continued to peel back the onion. But we also brought in my entire staff, so everyone could learn how AARs worked and could chime in with off-the-wall questions.

As we dug deeper and deeper, we discovered significant errors in the CIA's categorization and counting of terrorist incidents. These were errors, nothing more. They were not evidence of criminal, corrupt, or otherwise evil practices. The CIA's errors were then compounded by my staff, who had to admit they hadn't done an adequate job analyzing the draft report. The discussions were all conducted civilly and deliberately; no crucifixions were ordered.

By the third morning's AAR, everyone who knew anything about the issue was pitching in to make sure we had a clear view of exactly what had gone wrong. The AAR approach cut through all the Gordian knots and got to the core problems in short order. My staff and the CIA, working side by side, soon went to work redoing the analysis.

I called Waxman to tell him that he was right and I was wrong and to assure him that my team was hard at work fixing the problem and preparing an amended report. Because he trusted us, he gave us the time we needed. We submitted an accurate revised report within a few weeks. Congressman Waxman publicly congratulated us, and there was no further congressional intervention. More important, we fixed the report-making system to avoid future problems.

The problems I found were organizational and needed correction, and they were dealt with quietly and in a timely manner outside the AAR process. The goal of an AAR is to get everyone around a table to review the battle, learn what went wrong, learn what went right, and work out how to train to do better. Leadership and personnel problems revealed by AARs normally get fixed privately.

Every organization needs to be introspective, transparent, and honest with itself. This only works if everyone is unified on the goals and purpose of the organization and there is trust within the team. High-performing, successful organizations build cultures of introspection and trust and never lose sight of their purpose.

Chapter Twenty-Eight
Squirrels

One morning early in 1988, shortly after I became President Reagan's National Security Advisor, I went into the Oval Office to discuss a problem with the President. We were alone. He was sitting in his usual chair in front of the fireplace with a view of the Rose Garden through the beautiful glass-paned French doors. I was sitting on the end of the couch to his left.

I don't even remember what the problem was. But it involved a not-uncommon fight between the State and Defense departments, made more complicated by significant Commerce, Treasury Department, and congressional interests. I described the problem at some length and complexity to the President, underscoring that it had to be solved that day.

To my discomfort, he kept looking past me through the French doors without paying much attention to my tale of woe. So I talked a little louder and added more detail. Just as I was running out of gas, the President raised up and interrupted me: "Colin, Colin, the squirrels just came and picked up the nuts I put out there for them this morning." He then settled back into his chair and turned back to me. I decided the meeting was over, excused myself, and went back to my office down the hall in the northwest corner of the West Wing.

I had the feeling that something important had just happened. I sat down, gazed out my windows across the north lawn and into Lafayette Park across Pennsylvania Avenue, and reflected on it. And it became clear.

The President was teaching me: "Colin, I love you and I will sit here as long as you want me to, listening to your problem. Let me know when you bring me a problem I have to solve." I smiled at this new insight. In my remaining months with him, I told him about all the problems we were working on, but never asked him to solve problems that he had hired me and the rest of his team to solve. Reagan believed in delegating responsibility and authority and he trusted those who worked for him to do the right thing. He put enormous trust in his staff. The President's approach worked for him most of the time. But it could also get him in trouble, as the Iran-Contra debacle demonstrated.

On another morning in 1988, I went into the Oval Office with another problem. U.S. naval forces in the Persian Gulf were chasing Iranian gunboats that had threatened them. Our ships were approaching the Iranian twelve-mile limit and Secretary of Defense Frank Carlucci wanted authority to break that limit in hot pursuit of the boats.

President Reagan was sitting behind his desk, calmly signing photos, knowing that we were in action. He trusted our ability to manage the situation and keep him informed. He looked up as I approached and locked his eyes on me. He knew he was about to be handed a Commander in Chief problem to solve. I laid out the request, with all the upsides and downsides, potential consequences, press needs, and congressional briefing strategy. He took it all in and simply said, "Approved, do it." I conveyed the answer to Frank, we chased the boats back to their bases, and the action was over.

On many occasions during our time together, I brought Reagan presidential decisions that he would think through, question, analyze, and make. He was always available for Oval Office decisions. But he was happier if problems could be solved at a lower level.

One of my most treasured mementos, and the only signed picture I have from Reagan, shows us sitting side by side in front of the Oval Office fireplace. We are leaning toward each other examining charts I am using to explain some issue. He later inscribed that photo, "Dear Colin, If you say so, I know it must be right." Gulp.

I have always loved making things work well. From rebuilding worn-out Volvos to reshaping senior executive staffs into fine-tuned instruments, one of my deepest passions has been taking something that is not functioning as well as it should up to its highest level of performance. President Reagan taught me how to better achieve that goal by creating and maintaining mutual trust and accountability with my senior staff. They're as essential to a smooth running organization as an electrical system or driveshaft is to a Volvo.

In all my senior positions after serving under Reagan, I worked hard to create a Reaganesque level of mutual trust and accountability—trusting that my senior officials would be prepared, do the right thing, know what I wanted done, and be ready to be accountable for their actions.

Maintaining mutual trust and accountability meant keeping my people close to me, with very short and direct lines of communication and authority and the fewest possible bureaucratic layers between us.

My military training rested on the concept of the chain of command where everyone knows who is in charge and where only one person at a time can be in charge. From this training came my belief in working with direct-reporting subordinates without lots of assistants or other intervening layers helping me to run the staff.

My later experience in government has been that staff numbers multiply and fancy titles proliferate in an inexorable kudzu manner unless the bosses regularly and viciously prune them.

When Frank Carlucci was named National Security Advisor at the end of 1986 and asked me to be his deputy, I had only one condition. I wanted to be the only deputy: our predecessors had three. Frank readily agreed and the other two deputies were redesignated and given other duties.

I kept this model when I replaced Frank as the National Security Advisor, and followed the same model when I became Chairman of the Joint Chiefs of Staff in October 1989. My immediate line subordinates were two- and three-star generals and admirals, experienced officers at the top of their profession, each averaging more than twenty years of service.

At that time, the Chairman had an organization in the front office known as the Chairman's staff group. It consisted of four extremely talented colonels and Navy captains who monitored all the work coming in from my line subordinates. I eliminated the staff group. If a three-star general couldn't send me quality work, I didn't want a layer of bureaucracy covering for him. My now direct-reporting line subordinates quickly learned that once they signed something, I was the next stop. I was counting on them to make it right, and they knew it better be. Trust, responsibility, and accountability for results all go hand in hand.

When I became Secretary of State I put together a robust administrative office staff, but I neither wanted nor had special assistants in the front office working on substantive issues. That was the job of my senior line subordinates, the Assistant Secretaries who each presided over a large staff and had responsibility for the different regions of the world and major functional activities.

Historically, the Secretary of State had only one Deputy Secretary, whose chief function was to serve as an alter ego to the Secretary, and whose primary focus was on management issues. However, because Congress felt that the department was not being managed well, they had authorized a second Deputy Secretary of State to focus solely on management. I dutifully nominated someone for this statutory position; but, remembering the reasons for my demand to Carlucci, I saw to it that the nomination was never acted upon. I wanted only one deputy, Rich Armitage. Neither of us believed that management of the department was separate from our responsibility to manage foreign policy. Rich was more than capable of doing both, and we were blessed with talented subordinates who understood how we operated.

Another authorized position I didn't fill was called "Counselor of the Department," whose job was to do whatever the Secretary wanted him to do. It was a position I didn't need because my preference was always to use my line leaders and not supernumeraries.

A case can be made that my lean, direct-reporting approach was shortsighted, and that the Secretary of State should have constellations of special assistants orbiting around his front office to extend his personal presence. My successors felt the need to fill the second deputy position, they occasionally filled the counselor position, and they also added special emissaries in considerable numbers to oversee high-profile or sensitive issues of unique importance. Sometimes these emissaries added value, sometimes not. There is a place for special emissaries for selected missions, limited in time and scope, but foreign leaders have been confused at times about who is in charge of what. Special emissaries cannot substitute for the permanent staff.

I want to make it clear that my choices and the reasons I made them are not judgments on either my successors or predecessors; I am just saying that their preferences were not mine. Every senior leader up to and including the President must organize his team in a way that is consistent with his needs, experience, personality, and style. There is no single right answer to the "how do I organize my staff?" question.

I have always preferred to keep my staff at the top as small as possible and to work directly with my senior subordinates, whom I vested with authority and influence. Because I considered them an extension of me, they had to be close to me, and they had to know that I believed that when they said so, it must be right.

President Reagan taught me more about leadership than creating trust and accountability; he was an example of how the leader at the top has to step outside the pyramid of the organization to see the wider view from atop the highest hill of the shining city. He was always at what I call a higher level of aggregation than the rest of us.

One morning the President's entire economic team marched into the Oval Office to discuss a problem. The Japanese, who were then doing well economically, were buying up lots of U.S. properties—even icons of American real estate like Rockefeller Center and Pebble

Beach Resorts. Congress was starting to stir, the public was buzzing, and the purchases were causing economic and security concerns. "Something has to be done and done now," said the team. They quietly waited for the President to respond. He did.

"Well," he said, "they are investing in America, and I'm glad they know a good investment when they see one." The meeting was over. Reagan once again demonstrated his confidence in America. He was above our ground-level view.

Postscript: the Japanese greatly overpaid for their American investments. Rockefeller Center and Pebble Beach were soon back on the block; the Japanese lost money on the sales.

If Reagan were in the Oval Office today, he'd say the same thing about Chinese investments in America, and then go put out nuts for the squirrels.

Chapter Twenty-Nine
Meetings

When I was assistant to Secretary of Defense Cap Weinberger, one of my duties was to help the receptionists move about a dozen chairs into his office for the morning staff meeting. When it was over we moved the chairs back out.

The meeting was called the "LA/PA" meeting. Mr. Weinberger wanted to start the day hearing about hot media stories and Capitol Hill mischief from his legislative affairs and public affairs assistant secretaries. The other assistant secretaries of defense also attended, to listen to what was going on and to raise other pressing issues. The meeting lasted about thirty minutes. It was a very useful way to start the day.

Once or twice a month, Secretary Weinberger chaired a meeting of the Armed Forces Policy Council, a more formal body that assembled in his conference room. All the service secretaries, senior military leaders, and the Secretary's top staff members attended. This meeting had zero substance. It was absolutely useless. Well, not quite. The attendees could report to their staff and family that they had actually seen the Secretary that month. Because it was formal, infrequent, and had no real purpose, we had to struggle the day before the meeting to come up with issues for Weinberger to talk about. He barely scanned our paper before he headed in. During the meetings, people scribbled furiously as he droned on about the issues we had given him to drone on about.

Presidential cabinet meetings are no different. I attended cabinet meetings in four administrations. They were all the same. I'd be shocked to learn that they have changed.

For obvious reasons, they are not held on any particular schedule. The cabinet assembles in the Cabinet Room. Media come in, listen to the President discuss whatever subject interests them that day, and then leave. The President gives a pep talk to the cabinet. Designated cabinet members do a departmental show-and-tell or discuss a particular timely issue. People chat. After an hour they all depart. In the United States we don't really have cabinet government.

When I was National Security Advisor, Chairman of the Joint Chiefs of Staff, and Secretary of State, I used a variation of the Weinberger LA/PA model—an early morning meeting of my direct reports and principal aides that I called "Morning Prayers" . . . just starting the day together. It was a large meeting; as many as forty people attended. I had very strict rules:

"My morning meeting will never run longer than thirty minutes, usually less, so we can all get to work.

"This is the way we start the day as a team. I want you all to see me and check my morale and whether I seem okay. I want to look around the room at each of you and discern any subtle signals suggesting something I need to probe.

"This is not a show-and-tell meeting. If you have nothing to say, don't speak.

"No one gets reamed out here. We are sharing with each other, talking about the needs of the day and what we need to do, discussing how to fix problems. If anyone has really screwed up and needs counseling, we'll do it later, alone, in my office.

"You will leave the meeting knowing what is on my mind and, therefore, had better be on your mind. I want each of you to meet with your staff to share with them what we have discussed. We need to connect from top to bottom.

"And oh by the way, you can tell your spouse and relatives that you see the Secretary every day."

I told them not to be surprised if I poke fun at them or get a little goofy sometimes, too.

One Tuesday morning I came in and asked if anyone had seen *Monday Night Raw* wrestling with Hulk Hogan and the Undertaker. I was met with blank stares and bewilderment from the assembled ambassadors, senior Foreign Service officers, and other intellectual types. I described the match. It was a heavily choreographed ballet, I had to admit, yet the wrestlers showed considerable athletic skill and training as they bounced each other on the mat.

The bewilderment remained until I told them why the match interested me. Thirty thousand people had come out on a Monday night in a mid-sized city in the Midwest to watch it. This is what average Americans do. They also love NASCAR and Walmart. These are the folks we really work for. We can't forget that.

One of my jobs as Deputy National Security Advisor under Frank Carlucci was to convene interagency committees to resolve issues for the cabinet and the President. These meetings were more formal and serious than LA/PA-type meetings, with very senior attendees. Out of them came advice for presidential decisions. Thus their name—"Decision Meetings," because they had to end in a recommendation to take to the President for his decision. Attendance was at the deputy and undersecretary levels from the State, Defense, Treasury, and Commerce departments. The CIA was there, the Attorney General's office, the National Security Council staff expert, and White House experts. I had the Chair.

We had a lot going on in those days, with lots of meetings. We had to keep the trains moving on time. Meetings needed a tight structure.

An agenda was always set and briefing papers provided to each attendee well in advance. If you haven't had time to read the paper, send someone who has. Don't waste our time.

I would open the meeting with a five-minute description of the issue and the current state of play. For the next twenty-five minutes each agency with a position on the issue gave its presentation without interruption. For the next twenty minutes we had a food fight. Anyone could jump in and disagree and support or attack whoever they wanted. Strong language, passionate views, fight for your position. "Nothing personal, Sonny, just business."

At fifty minutes, I took over the meeting and made everyone shut up, summarizing the merits and demerits of the arguments and reaching a tentative conclusion that I would recommend to the President for his decision. This took five minutes. The attendees then got five minutes to object and clarify. If my recommendation still seemed right, I would confirm that. The meeting was over. They returned to their departments and briefed their Secretaries. If a Secretary strongly disagreed, he or she would call me that night. The next day the decision paper—with concurrences, nonconcurrences, and options—went to the President.

The paper reflected all the edges of our debate. Everyone present had their say. We didn't want to round the issue into a small beach pebble that might roll in any direction. Any cabinet Secretary who still strongly disagreed could go to the President.

I don't recall a single instance when a Secretary did that. We had made sure every view had been presented, considered, and reflected in the paper.

As with so many issues, we could often whittle a problem down to a series of alternatives, each of which should work. We then tried to pick the best of breed.

The full NSC worked the same way, with the President in the Chair. Cabinet officers presented their views, while either Frank or I served as the master of ceremonies, laying out the issue and guiding the discussion. The President usually asked questions, but seldom made a decision during the meeting. After it was over we prepared a decision paper for him. When he reached his decision, a written confirmation was disseminated.

Because some NSC staff members had gone rogue during the Iran-Contra period, we made sure the process was formal and documented. We succeeded in restoring credibility to the NSC system.

I've run many other kinds of meetings.

Informational or briefing meetings simply inform attendees about a subject of immediate interest. I kept a time limit on these meetings to keep them from wandering all over and to prevent spring butts from popping up mostly to hear themselves talking.

The principal official meeting of the Joint Chiefs of Staff was called a "Tank Meeting," so called because the original room where the World War II Joint Chiefs met was located in the basement of the Commerce Department and was reached through a tunnel-like entrance. In later years the meeting was normally held in a special room in the National Military Command Center at the Pentagon. The chiefs, their operations officers, and lots of backbenchers and note takers attended. There was a formal agenda.

I found it more useful to have the Chiefs meet in my office without assistants or agenda. This relieved them of their bureaucratic veneer, and we could talk openly—the most senior and experienced military guys in the armed forces and not just the leaders of large organizations whose interests the service Chiefs were expected to defend at all costs. Those meetings worked beautifully, as we dealt with the most fundamental issues of war and peace.

Not all my meetings were structured. I liked to end the day, for example, with a freewheeling get-together where three or four of my closest associates could sit around my office, feet up, and review how things were going. It was an unhurried time when we could prepare ourselves for the challenges and opportunities of the next day.

Humans are not by nature solitary. They need to connect with other human beings to share dreams and fears, to lean on each other, to enhance each other.

Two people together are a meeting. As organizations become larger, ever more people need to meet formally and informally. I have always tried to conduct meetings, no matter how large, with the intimacy and respect two longtime close friends show each other when reminiscing about their shared past.

Chapter Thirty
The Indispensable Person

During the worst days of the Civil War, President Lincoln would often get away from the summer heat of Washington by riding up to a telegraph office on a cool hill north of the city. The telegraph was the first great technology of the revolution in telecommunications that over time developed into communications satellites and the Internet. The President would sit in the telegraph office receiving the very latest reports from the battlefields.

One night a telegraph message came in detailing yet another Union army calamity. Confederate cavalry had surprised a Union camp near Manassas, Virginia, and captured a brigadier general and a hundred horses. With the telegraph operator watching, Lincoln slumped in his chair as he read of this latest setback.

Moaning slightly he said, "Sure hate to lose those one hundred horses."

The operator felt obliged to ask, "Mr. President, what about the brigadier general?"

Lincoln replied, "I can make a brigadier general in five minutes, but it is not easy to replace one hundred horses."

A friend gave me that quote in a frame the day I was promoted to brigadier general. I've made sure it was hanging right above my desk in every job since. My job as a leader was to take care of the horses, get the most out of them, and make sure they were all pulling in the direction I wanted to go. And, by the way, make sure there were folks behind me ready to be promoted to brigadier general and take over after I left.

"If I put you all in a plane and it crashed with no survivors," Army Chief of Staff General Bernie Rogers said to us during his welcoming speech to my class of fifty-nine new brigadier generals, "the next fifty-nine names on that list will be just as good as you. No problem."

During the run-up to Desert Storm, General Norm Schwarzkopf became ill. Norm was vital to the success of our plans, but I could not let him be indispensable. I had a replacement in mind, should that ever become necessary; and my boss, Defense Secretary Cheney, knew who I would recommend.

General Max Thurman, the commander of Southern Command from 1989 to 1991, planned, executed, and led our campaign in Panama to overthrow the dictator Manuel Noriega. Max was one of the greatest soldiers I've ever known and one of the dearest of friends. After the invasion was successfully completed, Max was diagnosed with cancer. During the early stages of treatment, he remained in charge of Southern Command, but after some months, it became clear that his treatment was going to be very intense and conflict with his duties. Secretary Cheney, who was close to Max, didn't want to relieve him. I finally persuaded him it was necessary. Max understood perfectly. In battle you take casualties and you move on. The needs of the mission and the horses must come first. Max eventually died of his illness.

Back when I was a young lieutenant in Germany, I was the executive officer of an infantry company, second in command to Captain Bill Louisell. We were out on a graded exercise testing our combat readiness—one of those exercises that tries to replicate as close as possible real combat conditions. On the second night, at the height of the action, the evaluators killed Louisell and removed him from the exercise. I was in command. We made it through the night and successfully completed the exercise. The credit went to Bill. He had kept me informed, trained me, and let me inside his concept and plan. I was able to take over when he was taken out.

I have run into too many people in public life who think they turn on the sun every morning. If not for them there would be no light and heat. I have run into too many people who have long passed their sell-by date and don't accept that it's time to leave. I have run into too many leaders who have never given a thought to succession or building a leadership team in depth. Too many leaders are too insecure to face those realities.

And I have run into too many leaders who would not face the reality that the indispensable person is holding their organization back. Leaders have an obligation to constantly examine their organization and prune those who are not performing. The good followers know who the underperformers are; they are waiting for a leader to do something about them.

When necessary pruning is not done, good followers often slack off. But when it is done successfully, black clouds lift from over the team.

Even the best, most treasured, most successful members of a team can lose their edge and become underproductive. Leaders need to be ready to replace anyone who is no longer up to the task. Don't reorganize around a weak follower. Retrain, move, or fire them. You are doing that person a favor in the long term. And you are doing your team a favor immediately.

Chapter Thirty-One
Time to Get off the Train

One of my dearest military friends, Colonel Frank Henry, was a fellow brigade commander in the 101st Airborne Division back in 1976. A great commander and as feisty as they come, Frank occasionally got in trouble crossing swords with our division commander.

We were talking one day about our career prospects. "I don't know if I'll go any higher in the Army," he told me; "but I'm proud I made colonel. The next thing I expect from the Army is to be told when it's time to get off the train."

I once shared that story with Larry King, the famous television host; he never forgot it. In 2010, his longtime CNN show, *Larry King Live*, was losing its audience. The information revolution was changing all media. It was becoming clear that CNN might terminate his show. Larry didn't wait. He made a sudden announcement that he would be stepping down after twenty-five years on the air. When he made the announcement he retold my old Frank Henry story. He'd had a great ride, he explained, but he'd reached his station. It was time to get off.

I tried to maintain the same attitude throughout my career. Working hard and leaving to the Army the decision about where to get off became a touchstone for me. They never made promises about how high I would go. "Just do your job well and you'll move up. We'll let you know when you have arrived at your station." I asked my conductor a number of times if the next station was mine. "Not yet," he kept telling me. And I kept riding.

My family was pleased that I'd gone into the Army. It was a patriotic duty and they loved our country. But for a long time they had trouble understanding why I stayed in. My aunt Laurice, the family doyenne, was assigned to press me on the issue when I returned from my second tour in Vietnam. Laurice was a master at getting in other people's business, and she was all over me. I finally got her off my case when I explained that if I worked hard I could retire as a lieutenant colonel with a 50 percent pension at age forty-one. For my immigrant family, a pension for life was the equivalent of a Powerball lottery hit. They never raised the question again.

I made lieutenant colonel. Everything after that became a frequent rider benefit and a blessing.

The Army has very strict up-or-out policies to keep the officer corps refreshed and to bring up young officers. I was honored and pleased in 1986 when I was selected for promotion to three stars, lieutenant general, to take command of the V Corps in Germany.

A letter from General John Wickham, the Army Chief of Staff and a longtime mentor, notified me of my promotion and new assignment, congratulated me, and ended with a notice that the assignment was for two years. After two years, *to the day*, if he hadn't selected me for another three-star position, or if I hadn't been selected for a fourth star, he expected that day to have my request for retirement on his desk. If I didn't he would be waiting at the station with one of those old mailbag hooks to yank me off.

I wasn't a corps commander long. After six months I was reassigned to the White House, first as Deputy National Security Advisor and then as National Security Advisor. These were positions of great responsibility and I was honored to be selected, but they badly mangled my military career pattern.

"We serve where we are needed and career progression be damned," General Wickham reminded me.

As President Reagan was leaving office, President-elect George H. W. Bush offered me several high-level positions in his new administration. I visited the new

Army Chief of Staff, General Carl Vuono, to get his advice.

"I've been away from the Army in nonmilitary jobs a lot in recent years, and I have options in the civilian world," I told him, "so I assume it's time for me to leave. But the Army is still my first love. I'd like to stay in the Army, but I'll accept any decision you make."

"The Army wants you to return," Vuono said, smiling, "and we're holding a four-star position for you." That was one of the happiest moments of my life.

When I told that to President Reagan the next day, he only asked, "Is it a promotion?"

"Yes," I answered.

"That's good," he said, in his simple, direct way.

President-elect Bush was gracious, but, I suspect relieved, since he now had another open seat in the first-class car he could fill with one of the many waiting in line for a job.

Over the years I've run into people who don't realize a station is waiting for them or who believe they have an unlimited-mileage ticket. Four-star generals with distinguished thirty-five-year careers have come into my office whining and pleading not to have to get off . . . as if they were entitled to stay on.

Presidential appointees at the State Department who had served for years at the pleasure of the President were appalled when I told them it was time to retire or move on to another job. One of them mounted a lobbying effort to suggest that I couldn't possibly do such a thing. I did it anyway. The wailing and gnashing of teeth was heard all over the department. That is, until the retirement ceremony was over and everyone else began to look at how their own career prospects had been affected.

Congress is probably the worst organization in this respect. I understand the importance of experience and the value of a decade or two of service. But thirty years or more? Give it up and give your great-grandson a chance. How many more federal buildings and roads do you need named after you!

No matter what your job, you are there to serve. It makes no difference if it is government, military, business, or any other endeavor. Go in with a commitment to selfless service, never selfish service. And cheerfully and with gratitude take your gold watch and plaque, get off the train before somebody throws you off, go sit in the shade with a drink, and take a look at the other tracks and the other trains out there. Spend a moment watching the old train disappear, then start a new journey on a new train.

Chapter Thirty-Two
Be Gone

Leaving the train is not just about when to get off, but how.

There's an old Army officer tradition. When you leave a post, you write "ppc" on the back of your business card and pin it to the officers' club bulletin board or similar public place. "Ppc" is an acronym for a French term *pour prendre conge*, in English, "to take leave." It was our final departing courtesy when we were making a Permanent Change of Station, with the emphasis on "permanent."

There's a more direct, colloquial way to put it: "When you're through pumping, let go of the handle."

I have seen too many executives in the private world hang on long after they have stepped down. They keep honorary, emeritus, or similar positions, which give them offices, assistants, and the ability to sit in and kibitz at meetings, enjoy the perks of office, and even receive compensation. Yet they have no responsibility or accountability.

In the Army, when it's time to go, you go. When you're an outgoing commander at a change-of-command ceremony, you get a medal, pass the colors to the new commander, give a short speech, and watch the troops march in review in your honor. You then shake hands with the new commander and walk off the field. If you do all this properly, your station wagon is behind the stands, all loaded up with suitcases strapped down on top. You, the kids, and your wife pile in, drive off, and head for the main gate while the new commander is going through his receiving line. It's important to make sure the car mirrors are oriented so you can't see behind you. It is even more important to keep the windows rolled up and the radio turned up high so you can't hear the trash can covers closing on all your great ideas. It's over. You've had your turn at bat.

For several months, people will call to tell you how much you are missed and how much trouble they're having with the new guy. It's all nonsense. Be patient. The calls will end before the new guy has finished heating up his branding iron.

I always tell my successor that I will never call him, but he should feel free to call me with questions. If he is newly promoted and we had worked together, my job was to train him to replace me. Now he has done that.

I detest long turnover periods. Study hard before showing up. Know everything you need to know before taking over. But keep the transition period short. Be polite and spend a little time with your predecessor, but don't overdo it. You really don't want to hear all about his tour and he probably resents you a bit.

When I took over from Madeleine Albright as Secretary of State, we met three times over a two-month period, once in her home and twice in her office. She was a good friend and I benefited from her insights. But otherwise, I stayed out of her way.

Once I took over Madeleine was always available, but she never called me or took issue with me in public. Four years later I had a similar be-gone experience with my successor, Condi Rice. After your turn at bat, head for the dugout, the bullpen, or the parking lot.

Most turnovers are nonhostile. But when a leader has been relieved for incompetence or misconduct, the situation is always tense. The new leader has to sweep clean quickly as he takes over, but he shouldn't beat up on the guy who got relieved. He has driven off post without a parade or medal. He knows what's happened. He's suffering for it. Don't beat up on his professional corpse.

PART VI
Reflections

Chapter Thirty-Three
The Powell Doctrine

You can't help but be flattered and honored to have a doctrine named after you. I haven't figured out yet how it happened to me.

The so-called Powell Doctrine exists in no military manual. The term first emerged in late 1990 after President Bush, on General Schwarzkopf's and my recommendation, decided to double the force facing the Iraqis. After Desert Storm the term entered the language, if not the manuals. It reflected my belief in using all the force necessary to achieve the kind of decisive and successful result that we had achieved in the invasion of Panama and in Operation Desert Storm.

In discussing what they take to be the doctrine's most essential element, commentators have tended to use the term "overwhelming force," but I have always preferred the term "decisive." A force that achieves a decisive result does not necessarily have to be overwhelming. Or, to put it another way, overwhelming force may be too much force. It's the successful outcome that's important, not how thoroughly you can bury your adversary or enemy.

I have always held the view that decisive force should be used in addressing a military conflict. The reason is simple: Why wouldn't you, if you could? After Desert Storm, during the question period following a speech to a naval audience at Annapolis, I was asked why I had sent General Schwarzkopf two additional aircraft carrier battle groups when he had only asked for one. My answer was simple: "I didn't have time to go get the rest of them. This is a gang fight." It was a great line, but my real reason was that I felt one more carrier than Norm requested added to the insurance policy that would give us ultimate victory.

The Powell Doctrine is often compared with the Weinberger Doctrine—six rules for the use of military force formulated by Defense Secretary Weinberger in 1984.

Though there are similarities between Secretary Weinberger's ideas and mine, I have never formally set down a list of rules. My views are not rules. I have always seen them as guidelines that senior leaders should consider as part of their decision-making process. The President decides if they are relevant or not to a particular situation. The military executes whatever the President decides.

My concept of the Powell Doctrine begins with the premise that war is to be avoided. Use all available political, diplomatic, economic, and financial means to try to solve the problem and achieve the political objective the President has established. At the same time, make it understood that military force exists to support diplomacy and take over where diplomacy leaves off. There is no sharp distinction between the two. Diplomacy that does not also imply the prospect of force may not be effective. If the readiness level of forces, deployments, exercises, and threats of use always remain on the table, we can often support diplomacy and achieve the President's political objective without firing a shot in anger.

But when the President decides that only force will accomplish his political objective, then force must be applied in a decisive manner. Without a clear political objective, you can't make an analysis of the required force.

In deciding what forces to use and how to apply them, planners must think the operation through in its entirety from start to finish. After you achieve your initial military objective, what then? How do you know when it is over, and how and why do you stay on or exit?

Later, as an operation unfolds, senior leaders must explain it to the American people and their representatives and to the rest of the world. Public support is not initially essential, but if you don't gain it over time, you will have trouble continuing the operation.

All of this assumes you have time to think, discuss, and plan. That doesn't always happen. Surprises happen. Crises suddenly erupt.

Presidents making critical decisions have to use all the information at their disposal, call on their instincts, and avoid being at the mercy of fixed guidelines or rules. The need to use force may come up urgently when, as the British say, "You are on the back foot." You have to act, decisively or not, clear political objective or not, public support or not. Those are the trying times when you earn your pay.

These principles apply all the way up through the largest, most complex military operations. But they rest on nothing more than fundamental principles of war that go back thousands of years. They could just as well have been called the Sun Tzu Doctrine or the Clausewitz Doctrine.

In the American Army, they are taught as Principles of War. I first learned them as an ROTC cadet. As currently taught, there are nine of them:

- Mass
- Objective
- Offensive
- Surprise
- Economy of force
- Maneuver
- Unity of command
- Security
- Simplicity

The first two are the classical formulation of the Powell Doctrine, with the order reversed. Here is how they are defined in Army manuals:

> Objective—Direct every military operation towards a clearly defined, decisive, and obtainable objective.
> Mass—Concentrate combat power at the decisive place and time.

Note the repeated *decisive*. Not only must mass be concentrated at the decisive place and time, but the objective must also be decisive. If you gain it, you win. (Clausewitz called this the "strategic center of gravity.")

When we launched the 1989 Panama invasion, our strategy was to take out not only the dictator, Manuel Noriega, but also his whole government and military force and replace them with a president already legally elected but in hiding. We used more than twenty-five thousand troops in a coup de main to quickly eliminate the Panamanian forces as a threat and consolidate our position. Then we shifted to protection of Panamanian society, installation of a new president and government, and reconstitution of the military. We were widely criticized around the world for unilaterally attacking a small country under what was believed to be insufficient provocation. The prompt and successful outcome of the operation quickly silenced criticisms. Today no American troops are in Panama, and in the two decades since the invasion the Panamanians have held four democratic elections.

President George H. W. Bush initially tried to counter the 1990 Iraqi invasion of Kuwait by means of economic sanctions and diplomacy, and he mobilized the entire international community to support these efforts. Our military mission during that period was to deploy troops to defend against the Iraqi army moving farther south into Saudi Arabia. We achieved that mission.

When it became clear that sanctions would not lead the Iraqis to withdraw, President Bush, at the recommendation of General Schwarzkopf and the Joint Chiefs of Staff, ordered a doubling of the size of the force in Saudi Arabia. The principal political and military mission of that force was clear—"eject the Iraqi army from Kuwait"—Desert Storm. We were certain the large force would be decisive, and so I was able to guarantee that outcome to the President.

That was the President's and all the planners' principal objective in Desert Storm.

Neither during the planning, nor during the actual operation, was there any consideration of marching on Baghdad, nor was there any political or international inclination to achieve that objective. We would not have had UN support; we would have been unable to build an international coalition; and President Bush had no desire to conquer a country. At the end of Desert Storm, the Iraqi army was no longer in Kuwait. Kuwait was firmly in the hands of its government. It was a military and political success.

At the very end of his term, President Bush sent troops into Somalia to restore order and permit the flow of food and other sustenance to a desperate people. The operation began in full view of television cameras. And of course the press made fun of the Navy SEALs seen wading ashore on every TV screen. So much for surprise. The truth was, we didn't want to surprise the ragtag irregulars who were making everyone else in that country miserable. We wanted them to see what was coming. We wanted them to be afraid of what we were laying down on them, and our visibility in the press helped us do that. In a few weeks, we accomplished the mission we'd set out to accomplish.

The incoming Clinton administration was determined to achieve a far more ambitious goal. They took on the task of creating a democracy where democracy had never existed and where there was never much appetite for it. After the tragic Black Hawk Down episode in October 1993 illustrated the futility of that effort, we pulled out of Somalia.

Later, in Bosnia, President Clinton got it right. The Serbian military was conducting violent and sometimes genocidal ethnic cleansing in that region of the former Yugoslavia. The situation was extremely complex; there was no clearly achievable political objective and no way to touch all the bases preferred by military doctrine. Even though the rules did not fit the situation, the President decided that action was required. Over a two-year period NATO slowly ratcheted up military operations against the Serbs. The President was right, and the NATO operations succeeded.

In September 1994, President Clinton decided that military force would be necessary to reinstall as the legitimate president of Haiti Jean-Bertrand Aristide, who had been removed from office by a coup led by General Raoul Cédras. As our troops were assembling and boarding planes, President Clinton dispatched former president Jimmy Carter, Senator Sam Nunn, and me to try to persuade General Cédras and the ruling military junta to step down. For two days we argued with Cédras and his generals.

At the critical meeting, President Carter asked me to explain what would happen to them if they didn't step down. I described the force that had been assembled and the tactics that would be used. "The force will arrive tomorrow," I told them.

Cédras gave me a long look. Finally, to break the tension, he said, "Hmm, Haiti used to have the smallest army in the Caribbean. Tomorrow we will have the largest."

Cédras and the junta saw it was time to fold their cards and leave the table.

When the 82nd Airborne arrived the next morning, they were greeted by General Cédras. After it was all over, I was reminded of one of my favorite classical maxims, sometimes attributed to Thucydides: "Of all manifestations of power, restraint impresses men most."

In their initial phases, the 2001 invasion of Afghanistan and the 2003 invasion of Iraq were extremely successful. Kabul and Baghdad fell quickly. We had taken out the governments, but the lack of clear and achievable follow-on objectives, or the means to achieve them, turned later phases into failures that took years and substantial surge forces to begin to reverse. The surge forces should have been there from the start. Wishful thinking had replaced strategic reality.

I could cite many more examples from American military history. And I could draw hundreds more from American corporate and political history—in fact, from just about any human endeavor.

Corporate leaders have to analyze their marketplace, their competitors, and the forces at their disposal. How do you mass your research and development, production, financial, and marketing forces to achieve your corporate goals? How do you deploy your leadership? How do you guard against surprises? When do you risk only using an economy of force? How do you exploit success or turn crisis and failure into an opportunity?

Even the Bible touches on these subjects. Luke 14:31 says:

> Or what king, going out to encounter another king in war, will not sit down first and deliberate whether he is able with ten thousand to meet him who comes against him with twenty thousand?

I would rather be the second king with twenty thousand than the first with his ten . . . and also have a clearer objective and a more decisive strategy.

Chapter Thirty-Four
The Pottery Barn Rule

I was taught as a young infantryman that after you've taken your objective, be it a hill, a town, a bridge, or a key road junction, you consolidate your position, get hot food and dry socks for the troops, bring up more ammo, dig in, and get ready for the counterattack. The battle ain't over yet and it will change form. As you consolidate your position and gauge the enemy's reaction, you look for opportunities to keep going. The enemy may have been so shattered that you can exploit your success and pursue him to final defeat. Or, he may have reinforced his forces and is coming back after you, perhaps in ways you hadn't thought of. Whatever happens next, you can be sure that you're going to face more action. Get ready for it. Take charge.

On the evening of August 5, 2002, President Bush and I met in his residence at the White House to discuss the pros and cons of the Iraq crisis. Momentum within the administration was building toward military action, and the President was increasingly inclined in that direction.

I wanted to make sure he understood that military action and its aftermath had serious consequences, many of which would be unforeseen, dangerous, and hard to control. Most of the briefings he had been receiving had been focused on the military option—defeat of the Iraqi army and bringing down Saddam Hussein and his regime. Not enough attention had been given either to nonmilitary options or the aftermath of a military conquest.

I had no doubt that our military would easily crush a smaller Iraqi army, much weakened by Desert Storm and the sanctions and other actions that came afterward. But I was concerned about the unpredictable consequences of war. According to plans being confidently put forward, Iraq was expected to somehow transform itself into a stable country with democratic leaders ninety days after we took Baghdad. I believed such hopes were unrealistic. I was sure we would be in for a longer struggle.

Wars break things, kill people, and leave in their wake horrendous confusion, chaos, and physical and social upheavals. Victory doesn't come automatically with the capture of the enemy capital. A defeated country under occupation is not a neat and orderly place. The old instruments of security and order are badly weakened or even totally destroyed. Normal transport and commerce are seriously disrupted. Even though the invading army may arrive as liberators, they may not be joyfully welcomed. There may be riots, looting, or widespread hostility to the occupiers, even sabotage and assassinations. Religious, political, or ethnic rivalries, kept under a lid before the invasion, may erupt unpredictably in the invasion's aftermath.

War is never a happy solution, but it may be the only solution. We must exhaustively explore other possible solutions before we make the choice for war. Every political and diplomatic effort should be made to avoid war while achieving your objective.

I had come up with a simple expression that summarized these ideas for the President: "If you break it, you own it." It was shorthand for the profound reality that if we take out another country's government by force, we instantly become the new government, responsible for governing the country and for the security of its people until we can turn all that over to a new, stable, and functioning government. We are now in charge. We have to be prepared to take charge.

After carefully listening to my presentation, the President asked for my recommendation. "We should take the problem to the United Nations," I told him. "Iraq is in violation of multiple UN resolutions. The UN is the legally aggrieved party. Let's see if there might be a diplomatic solution to the WMD issue. If not, and war becomes necessary, you will be in a better position to solicit the help of other nations to form a coalition.

"Of course," I added, "if the UN certifies to our satisfaction that there are no weapons of mass destruction in Iraq, that problem would be solved, but Saddam would still be in power. Is his elimination worth a war?"

The President and his national security staff, including Vice President Cheney and Secretary Rumsfeld, agreed with that course of action. The President presented it to the UN General Assembly on September 12, 2002, in his annual speech to the Assembly. In his speech he called for a Security Council resolution declaring Iraq in material breach of earlier UN resolutions and requiring that country to account for its WMD programs.

After eight weeks of intensive debate and negotiations, UN Resolution 1441 was passed unanimously by the Security Council. The resolution additionally held Iraq accountable for its dismal human rights record and its support of terrorism.

By early March 2003, the President and other world leaders decided that UN efforts would not succeed, and the war came. Military victory quickly followed. Baghdad fell on April 9, 2003, Hussein and his regime were brought down, we declared "Mission Accomplished" and celebrated victory . . . and chaos erupted. We did not assert control and authority over the country, especially Baghdad. We did not bring with us the capacity to impose our will. We did not take charge.

And Iraq did not in a few weeks magically transform itself into a stable nation with democratic leaders. Instead, a raging insurgency engulfed the country. Even as the country broke apart some senior members of the administration dismissed the insurgents as a few "dead enders," as if they would quickly fade away. They didn't fade away.

Three years later, the President realized the seriousness of the deteriorating situation and ordered a surge of troops to reverse the growing catastrophe.

The media gave a name to "if you break it, you own it": they called it the "Pottery Barn Rule." Though it did not come from me, the name was vivid, memorable, and accurately predictive, and the press stuck with it. The problem was that the Pottery Barn company had no such policy, and they were unhappy that people thought they did. And because the rule was associated with me, they were unhappy with me. Though I did my best to clear up the confusion in a television interview, the name stuck. Truth to tell, I wasn't sorry. Pottery Barn got spectacular publicity out of their nonpolicy.

Another truth: shops and commerce were never near my mind when I came up with "if you break it, you own it." For me the rule is all about personal responsibility; when you are in charge you have to take charge. The rule has nothing to do with Pottery Barn or any other store.

"Taking charge" is one of the first things a young Army recruit learns. The new soldier is taught how to pull guard duty—a mundane but essential task. Every recruit memorizes a set of rules describing how a guard performs his duty to standards. These rules are collectively known as the "General Orders."

One of those guard duty General Orders has stuck deeply in my head all these years and become a basic principle of my leadership style: A guard's responsibility is "to take charge of this post and all government property in view."

In other words, "When in charge, take charge."

Imagine an eighteen-year-old private walking around a motor pool where twenty tanks are parked side by side. It is cold, it is deep into the night, and he is alone. He is not just in charge of where he is standing or walking; he is in charge of this post. He is in charge of all government property in view, including the tank park, the buildings, his sleeping buddies, the fences, everything. Take charge doesn't mean keep an eye on it all or check occasionally. It means you are in charge, you are responsible, and you are supposed to act if anything is amiss or goes wrong. Even though you are only a private—as low as it gets in the Army—you are carrying the authority of your superiors and you had better act like it.

The General Order immediately following that one tells you to call someone whenever your instructions don't cover your situation. If you are confused or something weird happens, call for help. Call "the Corporal of the Guard." But until help arrives, you are in charge.

This establishes a "bias for action," a readiness to take on any challenge. We drill it into our officers and sergeants: "Don't stand there, do something!"

In the days, weeks, and months after the fall of Baghdad, we refused to react to what was happening before our eyes. We focused on expanding oil production, increasing electricity output, setting up a stock market, forming a new Iraqi government. These were all worth doing, but they had little meaning and were not achievable until we and the Iraqis took charge of this post and secured all property in view.

The Iraqis were glad to see Hussein gone. But they also had lives to live and families to take care of. The end of a monstrous regime didn't feed their kids; it didn't make it safe to cross town to get to a job. More than anything, Iraqis needed a sense of security and the knowledge that someone was in charge—someone in charge of keeping ministries from being burned down, museums from being looted, infrastructure from being destroyed, crime from exploding, and well-known sectarian differences from turning violent.

When we went in, we had a plan, which the President approved. We would not break up and disband the Iraqi army. We would use the reconstituted army with purged leadership to help us secure and maintain order throughout the country. We would dissolve the Baath party, the ruling political party, but we would not throw every party member out on the street. In Hussein's day, if you wanted to be a government official, a teacher, cop, or postal worker, you had to belong to the party. We were planning to eliminate top party leaders from positions of authority. But lower-level officials and workers had the education, skills, and training needed to run the country.

Yes, Iraq was a one-party tyranny. Yes, dangerous elements remained in the party and army. They would have to be identified and removed. Yes, many Iraqi soldiers had deserted. But viable structures remained, whose vacancies could easily be refilled.

The plan the President had approved was not implemented. Instead, Secretary Rumsfeld and Ambassador L. Paul Bremer, our man in charge in Iraq as head of the Coalition Provisional Authority, disbanded the Iraqi army and fired Baath party members, right down to teachers. We eliminated the very officials and institutions we should have been building on, and left thousands of the most highly skilled people in the country jobless and angry—prime recruits for insurgency.

These actions surprised the President, National Security Advisor Condi Rice, and me, but once they had been set in motion, the President felt he had to support Secretary Rumsfeld and Ambassador Bremer.

Meanwhile, at this decisive moment, we started sending troops home, removing senior commanders and their staffs, and cutting off the flow of additional troops. Back home we had "Mission Accomplished" celebrations, and the White House looked into arranging victory parades.

A hill had been taken, but the battle would continue for years to come. The victory over the Hussein regime was just the beginning of a long campaign, which we should have anticipated, but had not prepared for.

We broke it, we owned it, but we didn't take charge.

In 2006, President Bush ordered his now famous surge, and our forces, working with new Iraqi military and police forces, reversed the slide toward chaos. But years and many lives had been lost. U.S. and coalition forces have now left Iraq. Conditions in Iraq are vastly improved, but the campaign is still not over. We all hope that the Iraqis will be able to bring it to a successful conclusion and leave to their future generations a country free, democratic, and at peace with its neighbors and itself.

Any leader approaching an "if you break it, you own it" decision should preface his thinking with "try not to break it." But if there's a chance that you might break it, if you plan to break it, or if there's no way you can avoid breaking it, consider the costs of ownership and have plans ready to deal with the possible consequences of breaking it.

Plans are neither successful nor unsuccessful until they are executed. And the successful execution of a plan is more important than the plan itself. I was trained to expect a plan to need revision at the moment execution starts, and to always have a bunch of guys in a back room thinking about what could go right or wrong and making contingency plans to deal with either possibility.

The leader must be agile in thought and action. He must be ready to revise a plan, or dump it, if it isn't working or if new opportunities appear. Above all, the leader must never be blinded by the perceived brilliance of his plan or personal investment in it. The leader must watch the execution from beginning to end and do what it tells him.

Chapter Thirty-Five
February 5, 2003
The United Nations

Although it has been many years since I gave my famous—or infamous—Iraq WMD speech to the UN and the world, I am asked about it or read about it almost every day. February 5, 2003, the day of the speech, is as burned into my memory as my own birthday. The event will earn a prominent paragraph in my obituary.

"Is it a blot on your record?" Barbara Walters asked in my first major interview after leaving the State Department.

"Yes," I answered, "and there is nothing I can do about it."

What's done is done. It's over. I live with it.

Most people in public life have passed through a defining experience they'd prefer to forget, and to be forgotten, but won't be. So what can you do about it? How do you carry the burden?

In January 2003, as war with Iraq was approaching, President Bush felt we needed to present our case against Iraq to the public and the international community. By then, the President did not think war could be avoided. He had crossed the line in his own mind, even though the NSC had never met—and never would meet—to discuss the decision. On January 30, 2003, in the Oval Office, President Bush told me it was now time to present our case against Iraq to the United Nations.

The date he selected for the presentation was February 5, just a few days away.

The speech would cover several areas, from the Hussein regime's abysmal human rights record, to its violations of UN resolutions, to its support of terrorists. But its chief focus was to be its weapons of mass destruction. Though Saddam did not use WMDs during Desert Storm, he had them. He had used chemical weapons against his own people years earlier, and he had used them against the Iranians in the 1980–88 Iran-Iraq War. The intelligence community believed he not only still had WMD stockpiles, but also had continued to produce them. In the post-9/11 atmosphere there was deep concern that these weapons could get into the hands of terrorists.

Although the intelligence community differed about aspects of the Iraqi WMD program, there was no disagreement over the fact that the Iraqis had one. They were certain that Saddam had WMDs and was producing more. (UN weapons inspectors were always skeptical about these conclusions.)

Three months before my UN speech, the Director of Central Intelligence, at the request of Congress, had delivered to Congress a National Intelligence Estimate (NIE) that supported the intelligence community's judgment. Based on that NIE, Congress passed a resolution giving the President authority to take military action if the problem could not be solved peacefully through the UN.

The NIE contained a number of strong, definitive statements, including one claiming that Iraq was reconstituting its nuclear weapons program. In another: "Saddam probably has stocked at least 100 metric tons (MT) and possibly as much as 500 MT of CW [chemical warfare] agents—*much of it added in the last year* [my emphasis]." And in still another, the NIE claimed that the Iraqis had constructed mobile biological warfare production vans.

Though mostly circumstantial and inferential, the NIE's evidence was persuasive. It was accepted by our military commanders, the majority of Congress, the national security team, and the President, as well as by a number of our friends and allies. In the aftermath of 9/11, the President did not believe the nation could accept the risk of leaving a WMD capability in Saddam's hands.

Recognizing that we would eventually have to make our case to the international community, the President had directed the NSC staff to prepare that case.

Sometime later in the day of my January 30 meeting with the President, my staff received the WMD case the NSC staff had been working on. It was a disaster. It was incoherent. Assertions were made that either had no sourcing or no connection to the NIE. I asked George Tenet, Director of Central Intelligence (DCI), what had happened. He had nothing to do with it, he told me. He had provided the NIE and raw material to National Security Advisor Condoleezza Rice's office. He had no idea what happened to it after that.

I learned later that Scooter Libby, Vice President Cheney's chief of staff, had authored the unusable presentation, not the NSC staff. And several years after that, I learned from Dr. Rice that the idea of using Libby had come from the Vice President, who had persuaded the President to have Libby, a lawyer, write the "case" as a lawyer's brief and not as an intelligence assessment.

An intelligence assessment presents the evidence and the conclusions drawn from the evidence. A lawyer's brief argues guilt or innocence. Our biggest problem with the Libby presentation was that we couldn't track the facts and assertions with the NIE or other intelligence. The DCI could not stand behind it, and it was therefore worthless to us.

There was no way we could use the presentation as it came to us, and we had roughly four days to redo it. I asked for a delay, but the President had already publicly announced the date for my speech, and the UN had put it on the calendar.

"Okay," I thought. "we can handle that." I was disturbed, but not deeply troubled. We weren't working from scratch; we had the NIE and the CIA's raw material to draw from. On the other hand, our case had to be airtight. We were facing a moment like Adlai Stevenson's UN speech during the 1962 Cuban Missile Crisis, when he demonstrated to the world that the Soviets were beyond doubt installing nuclear-capable missiles in Cuba.

My staff moved to the CIA to work with Director Tenet, his deputy John McLaughlin, and their analysts. They worked for four days and four nights. Every night Dr. Rice, other White House officials, and I joined the group. The conference room was packed. We spent hours going over every detail, trying to come up with solid evidence, discarding any item that seemed a stretch or wasn't multisourced. Some items that I had to reject came from the Vice President, who urged us to tilt our presentation back toward Scooter Libby's by adding assertions that had been rejected months earlier to links between Iraq and 9/11 and other terrorist acts. These assertions weren't backed by what the intelligence community believed and stood behind.

The presentation was finalized at our mission in New York the night before I was to present it to the Security Council. My staff worked on it well into the night, and Tenet and McLaughlin stood by every word.

The next morning at the Security Council, I spoke for an hour and a half, broadcast live throughout the world. George sat right behind me. Though I wouldn't call it an Adlai Stevenson moment, my feeling was that the presentation went well. The British and Spanish foreign ministers supported us; the French foreign minister opposed us . . . about what we expected. On balance, we seemed to have made a powerful case.

The war began six weeks later, and Baghdad fell to our forces on April 9. In the first weeks no WMDs were found. In the weeks to come, hundreds of inspectors scoured Iraq. Scattered pieces of WMD-related debris were discovered, but no working WMDs were found.

As the world knows, no WMDs were ever found. There were none.

Although he retained the capability to start up again, Hussein had no existing WMD capability. (Predictably, conspiracy theories claimed that he had had his WMDs buried or sent to Syria. Those theories were baseless.)

Example: the biological vans reported by the CIA. At one point a van was discovered and photographed that seemed to fit the description of the biological vans the CIA had reported. President Bush quickly claimed the photos proved our case. But when my State Department intelligence staff examined them, they concluded that the vans were not bio labs. I agreed. The van that was photographed was crude, open, and poorly constructed; it only vaguely resembled a sophisticated facility for producing biological weapons. This was the closest anybody in Iraq came to finding WMDs.

Even though it was obvious to me and my staff that there was no way the vans could have produced biological weapons, a month after we got the photos, the CIA published a twenty-eight-page pamphlet insisting that was what they were for.

Over the following weeks, snippets of information from the CIA were briefed to the President and then to me that totally destroyed the credibility of other sources the CIA had claimed were solid. I was bewildered. How could we have been so far off the mark? How could our seemingly solid case have so devastatingly unraveled? Where could the NIE judgment have come from that the Iraqis had hundreds of tons of chemical warfare agents, "*much of it added in the last year*"?

In August, four months after the fall of Baghdad, even as their sources collapsed and no WMDs had been found, the CIA continued to formally report that based on what they knew and believed at the time they were made, they stood by their original judgments. In its findings the Iraq Intelligence Commission, created by presidential executive order and led by former senator Chuck Robb and Judge Laurence Silberman, detailed the intelligence community's failures in analysis and judgment. It was one of the worst intelligence failures in U.S. history.

Everyone remembers my UN presentation. It had enormous impact and influence in this country and worldwide. It convinced many people that we were on the right course. Members of Congress told me that I had persuaded them to vote for a resolution supporting the President—even though they had voted for that resolution three months before I spoke to the UN. My presentation became *the* case against Iraq. Who remembers any other?

Yet seldom is it mentioned that every senior U.S. official would have made the exact same case, or that many of them were in fact making that case on television and in other public appearances. We had all been convinced by the same evidence. None of us knew that much of the evidence was wrong.

If we had known there were no WMDs, there would have been no war.

Because the case against Iraq has become identified in so many minds with my UN speech, I still get asked about it frequently, and it's a target for regular attacks on the Internet. Were we lying? Did we know the evidence was false?

The answer to these questions is no.

There are other questions: Why did so many senior people fall for such shaky sources? Why and how did the CIA fail so massively? Did analysts decide to tell us what they thought we wanted to hear? It was even possible that we had been tricked by Iraqi disinformation. If Saddam wanted us to believe he had WMDs, then he convinced us.

I have no answers to those questions. I wish I did.

My questions don't stop there. I've asked myself again and again: "Should I have seen the NIE's weaknesses? Should I have sniffed them out? Did my critical instincts fail me?"

And then I read articles and books by former CIA officials describing their shock at the unsupported claims in my UN speech. Where were they when the NIE was being prepared months earlier, or when these same claims were being written into the President's January 2003 State of the Union address?

Yes, I was annoyed, and I'm still annoyed. And yes, I wish there weren't so many unanswered questions. And yes, I get mad when bloggers accuse me of lying—of knowing the information was false. I didn't. And yes, a blot, a failure, will always be attached to me and my UN presentation. But I am mad mostly at myself for not having smelled the problem. My instincts failed me.

Perhaps if we had more than four days, the weaknesses would have been uncovered. Maybe not; the intelligence community was telling me what they believed was known.

But I knew that I had to put aside my annoyance, distress, and disappointment. I knew that I had to live with the blot.

I was still the Secretary of State with a full plate. I had to shake this burden off, get on with my work, and learn from the experience. I learned to be more demanding of intelligence analysts. I learned to sharpen my natural skepticism toward apparently all-knowing experts.

I have never before written my account of the events surrounding my 2003 UN speech. I'll probably never write another.

It was by no means my first, but it was one of my most momentous failures, the one with the widest-ranging impact. And yet it was like all the others in this one respect. I try to deal with them all the same way. I try to follow these guidelines:

Always try to get over failure quickly. Learn from it. Study how you contributed to it. If you are responsible for it, own up to it. Though others may have greater responsibility for it than you do, don't look for that as an escape hatch. Once you have analyzed what went wrong and what you did wrong, internalize the lessons and then move on. As always, drive through life looking through the front windshield and not the rearview mirror. Don't become one of those pests who can't stop talking about their by now ancient slights, betrayals, hurts, or disasters. Don't wallow with your sympathetic friends. Learn and move on.

I am glad Saddam Hussein was removed from power. If he had escaped judgment in 2003 and got free of UN sanctions, I have no doubt he would have gone back to developing and producing weapons of mass destruction. That threat is gone. I admire the dedication of our troops and those of our coalition partners who fought the battles and are now home. I share a soldier's grief and sympathy for those who made the supreme sacrifice and for those who were wounded and scarred. And for their families.

As we move on, we must make sure the lessons learned are never forgotten or ignored.

Chapter Thirty-Six
Parsley Island

Leaders must be problem solvers. If you are not solving problems, you are no longer leading. Hopefully the problems you are solving relate to you, your organization, or your own interests. That is not always the case. Sometimes a problem comes totally out of the blue. It is of no interest to you, you have no skin in the game, you don't know the first thing about it, and yet you have to take it on.

Out-of-the-blue problem solving can grow even more complex if your organization happens to be the United States government, which—for better or worse—has long been the world's go-to problem solver.

On a quiet Thursday afternoon in July 2002, I received a phone call from the new Spanish foreign minister, Ana Palacio, in office for just a few days. I managed only a couple of congratulatory words before she got to the reason for her call. "We have a crisis in the Mediterranean," she said excitedly, "and you need to do something about it."

I had no idea what she was talking about, but rather than look like an idiot, I bought time. "I've been getting updated on the situation; let me call you back in a few minutes."

I put down the phone and screamed at the staff in the outer office, "What crisis in the Mediterranean? Haven't I told you about 'telling me early' and 'no surprises'? Is there a war going on that I don't know about?"

The staff called our resident European and African experts, who came charging into my office. "Mr. Secretary, here's what's going on. There's an island two hundred meters off the coast of Morocco named Perejil—'parsley' in Spanish. In English we usually just call it 'Parsley Island.' Perejil belongs to Spain, and has for four hundred years. Morocco disputes that, as well as Spain's ownership of two other enclaves on the coast of Morocco, Ceuta and Melilla."

"I never heard of the place," I replied. "I thought I knew the Mediterranean."

"Well, sir, it's a tiny, rocky outcropping about the size of a football field. Nothing much grows there except parsley, and there are no inhabitants other than feral goats. Sunbathers and drug runners occasionally stay overnight."

"Yeah, yeah, okay, so why do I have a crisis?"

"Well, sir, we have just had the first invasion of Europe from Africa since World War II. For reasons that aren't clear, the Moroccans decided to seize the island, perhaps to celebrate the king's recent wedding. The invading force consisted of a dozen Moroccan frontier guards who paddled across and put up a tent and two Moroccan flags. And they had a radio."

"All right, then what happened?"

"Well, a couple of days later, the Spaniards noticed they had lost their island, and all hell broke loose. It became a political crisis in Spain. The Spanish government notified NATO and the European Union. NATO punted; told them it was a bilateral problem. But the EU condemned the invasion. 'This is clearly a regrettable incident,' they announced. 'It constitutes a violation of Spanish territory.' The Moroccans took the issue to the Organization of the Islamic Conference (OIC) and got their support. No surprises there."

My guys continued: "Well, then the Spaniards attacked with naval forces, retook the island, and put the Moroccans back on their own beach. There are now seventy-five Spanish Legionnaires on the rock."

I had to smile. "Are you all pulling my leg? Isn't this a scene from *The Mouse that Roared*"?—alluding to the classic Peter Sellers comedy about a minuscule European country that gets hold of a superweapon by mistake and makes the great powers tremble.

"No sir, it has become a serious international issue."

I wondered, Why is Ana calling me? I was afraid to answer, but I had to call her back.

When I got Ana on the phone, I explained that I was now fully up to date on the crisis. "How could I be of help?" I asked reluctantly.

"Well," she replied, "we've got our island back; now our Legionnaires want to come home. But the Moroccans are waiting on the beach and might try to retake the island. The OIC supports them; the EU supports us; and so you have to solve it."

Bingo, I've got the brass ring.

Fortunately, no one was hurt in the invasion or the counterattack. When the Legionnaires arrived, only six Moroccans actually remained to garrison the island. The Legionnaires escorted them back to Morocco.

The solution was obvious: go back to the status quo ante bellum, the way it had been for four hundred years. It sounded simple to do.

Over the next forty-eight hours I made multiple calls to Ana Palacio and to the Moroccan foreign minister,

Mohamed Benaissa, a distinguished diplomat I had known for years. All kinds of arguments surfaced, but we successfully buried them. Finally, on Saturday morning we had a deal (I was now making all these arrangements from my home by telephone). We agreed that the Legionnaires would leave the island by 11:30 my time, a few hours hence. I was congratulating the two foreign ministers when they suddenly demanded that the deal had to have a written agreement.

"Go write one," I suggested.

Nope, I had to do it.

"Me? But who will sign it?"

"Easy, we want you to sign it."

They expected me to write and sign an international binding agreement between two foreign countries? It's a good thing I was home and my lawyers weren't around.

I went to work on my home computer. About ten minutes later, I had knocked out a one-page agreement. I faxed it to them, and more arguments broke out. The biggest was over the name of the island. Morocco objected to the Spanish name, Perejil, and the Spaniards wouldn't accept Leila, the preferred Moroccan name.

Hmmm. I called down to the State Department operations center. "Find our cartographers and get me the latitude and longitude of the crummy rock down to the minute and second."

They couldn't argue over that, and the two foreign ministers agreed to the document about a place with no name.

The Spaniards took the agreement to Prime Minister José María Aznar, and they briefed King Juan Carlos. They both okayed it.

But there was a hang-up at the other end. According to Minister Benaissa, King Mohammed VI was in a car in the desert and couldn't be reached. They couldn't approve the document until he had seen it.

It was now getting dark over the island. We only had thirty minutes to safely execute the departure of the Legionnaires. If they couldn't leave on time, the entire agreement I had cobbled together might fall apart.

"Time is of the essence," I told Benaissa. "I have other things to do," such as playing in our pool with my grandsons, Jeffrey and Bryan, who were about to arrive. "Don't know how you are going to do it, but I need to speak to the king in the next ten minutes." I had known the king for a number of years and had been close to his late father. I could take a few liberties.

Five minutes later the phone rang; His Majesty was on the line. I explained the essence of the document and made clear the urgent need for his approval.

"I can't approve it without studying it," he told me. "And I don't have a copy."

"Time doesn't permit that," I countered respectfully. "Your Majesty," I continued, "the United States and Morocco have been friends for over two hundred years. We would never knowingly do anything against your interest, nor would we do that to our other friend and ally, Spain. Sir, just trust us."

He paused for a moment, then announced, "Mr. Secretary, I approve. We trust America."

I thanked him, hung up, printed the document, signed it, and faxed it to Madrid and Rabat. The Legionnaires left shortly thereafter, and the Moroccans stayed on the beach. Ana went to Rabat a few weeks later for a lunch and conference with Benaissa, and all has been well ever since, at least with respect to Parsley Island.

The United States is the necessary nation. Despite our own problems, mistakes, and malfunctions, the world continues to look to us to solve or help with problems and crises, big and not so big, whether we have an equity in them or not. We are trusted. We are trusted to fight aggression, to relieve suffering, to serve as an inspiration to freedom-seeking people, to stand alongside our friends, and to welcome the tired, the poor, the huddled masses of other lands yearning to breathe free. That is who we have been, now are, and always must be.

After it was all over, Prime Minister Aznar called to thank me. "I'm thinking of vacationing next weekend with my family on Perejil."

I reminded him that the United States Navy still had ships in the western Mediterranean. We had a good laugh. On Monday, my lawyers were not too happy.

Ana and I became the best of friends.

Chapter Thirty-Seven
Pizza and Milk

Student exchange programs are wonderful things. Sending young Americans overseas, if only for a few days, opens their eyes to new experiences and gives them an understanding of a world that is not America—and a greater appreciation of what it means to be a citizen of this country.

Exchanges in our direction are no less important.

Bringing youngsters from around the world to the United States lets them experience the real America and the great people who live here. They'll see an America they'll never see on a screen.

The late Robin Cook was the United Kingdom foreign secretary during my first months as Secretary of State. In 1997 I founded a program called America's Promise, which helps youngsters in need get mentoring, safe places, and education.

Inspired by that program, Robin suggested that we exchange high school age kids between our two offices. I would find two young Americans to spend time with the Foreign and Commonwealth Office in London and he would send a pair of young Brits to hang out with my staff. It worked great, and it was fun for me to check on how the kids were managing to wrap their brains around our vast, complex, and strange organization. It was especially fun on their last day, when I'd bring the youngsters into my office and let them phone their "mums" back in the United Kingdom. After they and their parents had rattled on awhile, I'd take the phone for a minute and chat with the parents, which always pleased everybody.

The program continued under Jack Straw, Robin's successor. By then a way to improve it had come to me. "We both go out of our way to pick kids who are overachievers and professional winners," I told him. "Maybe you could send a couple who are not on their way to Oxford or Cambridge?"

He got it! Did he ever! He sent two young men who were not college-bound and came with all kinds of troubles behind them. They'd had run-ins with the law. They'd been busted for drugs. Their dress was not Savile Row, but council housing (public housing).

During their two-week stay, they met important people, visited our monuments, and spent a day with me. I took them to meetings at State and even to a congressional hearing. They saw what a Secretary of State actually does for a living. That afternoon I took them to the White House and we wandered around the fabulous eighteen acres. When we got to the Rose Garden I suggested that we check to see if the President was in the Oval Office. If he was out, we could see it. Surprise, surprise—I had earlier called President Bush and explained that I was bringing these boys with their troubled past to the White House. I knew he would be in.

We walked past the receptionists and right into the Oval Office, and there was President Bush waiting for us. The boys were amazed. There was small talk to break the ice. And then in a moment I will never forget, President Bush talked briefly but openly about his own onetime alcohol addiction and about how he had overcome it and gone on to create a new life, which eventually led to the Oval Office. After we left, I took the two speechless young men back to my office. Their lives were changed. Back in Britain they spread the word about this marvelous experience and the wonderful, kind, and generous people they had met.

The State Department has several youth exchange programs. One of them, Youth Ambassadors (YA), began in Brazil and was then exported to Argentina, Chile, Paraguay, Uruguay, and throughout the region.

High school students come to the United States for a few weeks, meet important people, see the sights, and take their impressions home.

In the winter of 2002, I received in my office a group of Brazilian YA students. We had a nice chat, but I could tell they were edgy. It had started to snow outside . . . their first ever snowfall. Since getting outside into the snow obviously seemed far more exciting than spending time with me, I let them go early.

Maybe six months later, I made a trip to Brazil. Curious about how the program had affected those kids, I asked our ambassador, John Danilovich, a great guy, to round them up so I could chat with them.

John located them, and we assembled in the backyard of his residence. The kids, sitting in a semicircle facing us, filled us in on their lives and their plans for the future. Since they were a highly selective group, they had expectations of future success—like owning companies or becoming their country's president.

I asked them if they had enjoyed the United States. I wanted to know specifically if anything had surprised them, or made them especially happy or especially sad.

There were questioning looks, but that didn't last long—they were teenagers.

One young man raised his hand. "One day we were having lunch at a school," he said, "and I was surprised, very surprised, when the American students laughed at me because I put ketchup on my pizza."

"Most Americans think pizza comes with a sufficient quantity of tomato paste," I explained kindly, doing my best not to smile.

Another young man quickly followed up. "I couldn't believe," he said with an expression of comic disgust, "that they served milk with pizza."

I again suppressed a smile. Time didn't permit an exposition about the place of the dairy lobby in the American political system.

Then a young girl tentatively raised her hand, "Let me tell you what happened to us in Chicago," she said.

"Uh oh," I thought.

"After a day of sightseeing, we went to a neighborhood restaurant," she explained. "I think it was an Outback Steakhouse. After we ate and the check came, we added up our money. We were short. We weren't used to paying with dollars. We couldn't pay the bill."

There they were, a dozen unchaperoned Portuguese-speaking kids in a chain restaurant in Chicago imagining all the horrors that could fall on foreigners who can't pay. When the waitress came back, the kids told her they couldn't cover the check. She looked at them, gave a nod, and went away. They didn't know what to expect.

A few minutes later, she came back. "Don't worry about the check," she told them with a warm smile.

"Will you have to make up the difference?" they asked, worried.

"Oh, no," she said, her smile broadening. "When I told the manager about your problem, he picked up the whole bill and gave me a message for you: 'I'm glad you came to our restaurant and hope you enjoyed the meal. I'm glad you're in our city and hope you enjoy your stay in America.' "

They were stunned. They never expected such kindness.

When the girl ended her story, the others remained silent. It had been a powerful experience for all of them. We had introduced them to congressmen, cabinet Secretaries, and other dignitaries, but a restaurant manager in Chicago made the strongest impression on them and gave them their most enduring memory of America.

Another young lady raised her hand. "We were boarding the plane to leave Chicago," she said. "After I sat down, a woman got in the seat next to me. 'Excuse me,' she said. I was confused. 'Why?' I asked. 'Well,' she said, 'I brushed against you when I took my seat. I hope I didn't disturb you.'

"I'll never forget that," she concluded.

A simple courtesy that most of us would have forgotten before the beverage cart rolled down the aisle left an indelible impression on a young Brazilian girl. It's hard to say why. Maybe she didn't expect such obvious niceness here, or maybe she wasn't used to that kind of gesture in Brazil. Whatever the reason, the moment has stayed with her.

When they returned home, the YA alumni appeared on Brazilian media and became multipliers of goodwill to the Brazilian people, especially to young people. There are now alumni in every Brazilian state.

None of the YA students have so far become corporate big shots or their country's president. But a few, like Casio, stand out.

When Casio returned to his small town, he decided to share his experience. "I realized the secret of my success was my mastery of English," he told John Danilovich. So he started his own language school, called Backpack. "Branding is important," Casio said. "You have to have a name they will remember." He marketed his school with his own website, and then went to the mayor of his town. "I'm going to start a language school that will help our town's young people," he told the mayor. "You should give me books for them." The mayor gave him books.

When Casio told this story to Ambassador Danilovich, John realized that the embassy could help, too, and they gave him books.

Later, YA alumni who got into the University of Brasilia began an entry exam preparation program run by Casio to help economically disadvantaged students prep for the rigorous entry exam. They charged ten reals, or about four dollars, for a semester of classes. "You cannot give it to them for free," Casio explained. "They won't appreciate it if it's free." Casio will have a brilliant future in marketing.

The YA success is a State Department success (it breeds lots of goodwill), but it's much more than that. It's an American people success. Our own people are our best ambassadors and promoters.

You never can tell what kids are really seeing (much less control it), but they are always seeing, and always judging. If we can provide them with rich enough experiences, they'll take something good away with them that they can use to make their own and other people's lives better.

Chapter Thirty-Eight
Cousin Di

My parents were proud British subjects. Although they became American citizens and loved their new country from the depth of their hearts, their Jamaican roots and their original British passports never let them forget their home. I was born in New York, yet I inherited their feelings about home and considered myself not only Jamaican but just a bit British.

I was given a very British name, Colin, pronounced "Cah-lin" by Brits and Jamaicans. In my youth in the early days of World War II, an American B-17 bomber pilot named Captain Colin—"Coh-lin"—Kelly heroically and successfully attacked a Japanese warship. His plane was severely damaged by Japanese fighters, but he held on until six of his crew members could bail out. The plane then exploded, killing Captain Kelly. He was one of the first American heroes of World War II. My friends started calling me by this Irish variant. No one cared until I became National Security Advisor, and the press demanded to know how to pronounce my name. I answered, "Coh-lin," to the dismay of my family.

British West Indians are proud of their heritage and Commonwealth connections. They also kid each other. My Jamaican family used to laugh over the message that tiny Barbados supposedly sent to King George VI at the beginning of World War II: "Carry on, England, Barbados is behind you."

It was many years before I was returned to my British roots. After the First Gulf War, in which the United Kingdom played an important role, Her Majesty's government saw fit to award me an honorary knighthood as a Knight Commander of the Order of the Bath. Because I am not a Brit but a citizen of a once rebellious colony, it was only honorary and had to be presented in a modest manner.

On December 15, 1993, Alma and I arrived at Buckingham Palace for the ceremony, which was to be hosted by Her Majesty Queen Elizabeth II. We were instructed by the Equerry that we would be announced into Her Majesty's office, she would make the presentation, and she might or might not choose to invite us to sit and chat. She would be alone in the room; there would not even be a photographer.

At the appointed moment, we entered the queen's small, elegant office. As she walked across the room toward us, she passed by a small table and picked up a leather box with the award inside, and approached us. "How nice to see you again, General and Mrs. Powell," she said, then added, "I'm pleased to give you this," and handed me the box. No pomp, no sword, no ermine robe, no photographer. She then invited us to chat, and we had a lovely fifteen minutes. Alma and I would enjoy her gracious company a number of times in the years ahead.

After leaving the palace, we posed outside for a photo and stepped into the marvelous Rolls-Royce limousine provided for us by the Foreign and Commonwealth Office. The liveried driver looked over his shoulder and said to Alma, "And where would you like to go now, Lady Powell."

"To Harrods, my good man," she replied with a royal smile. And she's never been the same.

We were privileged over the years to meet other members of the royal family. All of them were memorable, but Princess Diana was the most memorable.

We first met her in October 1994, at a luncheon in her honor at the British embassy in Washington. She was every bit as lovely in person as in her photos. We got along splendidly. I suspect the British ambassador had assured her that because of our military penchant for secrecy, she could relax and need not be guarded in her conversation with us. And she wasn't. Neither of us ever broke that confidence.

About that time a London newspaper wrote an article suggesting that Princess Diana and I shared a genealogy that could be traced back to the Earl of Coote, who lived in the 1500s. Though that seemed a stretch, I pocketed the news immediately.

We met again in 1995 in New York at a charity fundraising dinner for cerebral palsy research. It was an A-list, black-tie event where both of us were being honored. Barbara Walters was to introduce me and present my award. Henry Kissinger would then do likewise for Her Royal Highness. Needless to say, I was the second banana in this act, and Henry was in seventh heaven—the envy of every man in the room. Standing in the receiving line next to Diana, I got a sense of how hard it must be for her to endure the smothering public life she led. I almost tossed one guy out of the line when he shoved himself between us, draped his arm around her, and shot a self-portrait with a pocket camera.

After dinner came the presentations. I was first, and I wanted to put Henry, a beloved old colleague, off his pace (for fun). Barbara made the presentation, then I took the lectern, thanked the sponsors, praised the charity, and closed by announcing how especially humbled I was to be sharing the honors with Her Royal Highness, "with whom I had a relationship." The room was silent for a beat. Then came a small, general gasp. Alma shot me a wife look.

I sat down and Henry took the stage, a little off balance. But pro that he is, he recovered and gave Diana a splendid introduction.

Her remarks began, "Dr. Kissinger, ladies and gentlemen and Cousin Colin, good evening." Match point, Henry!

But the fun did not last. A few minutes into her speech came another incident highlighting the terrible demands of celebrity. A woman in the audience shouted out, "Why aren't you home with your children?" Everyone was stunned, but Diana didn't miss a beat, saying, "They are just fine, thank you very much," to much applause. I only hope that the anonymous doyenne raised her children as well as Diana raised William and Harry.

It was a year later at another black-tie charity event that we really became friends, this time at a dinner-dance for breast cancer research in Washington. Earlier that year, she'd been in Chicago for another black-tie dinner-dance. Before the event, a stalker had sent an incredible profusion of flowers to her hotel suite; when the dancing began, the stalker had managed to get into the queue and dance with her. Scotland Yard security was not happy, and since they did not want a repeat, there would be no strangers in the queue for the Washington dance. I was asked to be the first gentleman to dance with her. I would be followed by Oscar de la Renta, and other New York fashionistas. Well, it was tough duty, but someone had to do it.

At a British embassy luncheon earlier that day, Diana and I sat next to each other; one of the topics we touched on as we chatted was the dance that evening. After lunch she suggested that we practice a little. That seemed sensible, so we danced, without music, in a room next to the embassy dining room. When I asked about music that evening, she told me that any would do, but she offered one caution. Her dress for the event was backless; I would have to decide where to place my hand. I thought I could handle that, and raced out to buy new shoes. The evening was a tremendous success, and guys were staring daggers of envy at me.

In the years that followed we exchanged Christmas cards and an occasional letter until the terrible night in Paris when she was killed.

The celebrity of her position as the People's Princess created the conditions that led to her death. Paparazzi, tabloids, the expansion of the Internet, the explosion of social networks, and the introduction of cameras into phones and ever smaller cases make everyone in public life much more vulnerable. Intrusions by the media are no longer an occasional irritation; they're constant. All of this feeds an insatiable, often vicious appetite for the celebrification of our society. The more outrageous, misanthropic, and narcissistic the behavior, the more it sells. We suck it all up. The news and gossip cycles now move so fast that a falsehood goes around the world at the speed of light and is embedded in a million depositories. The correcting truth seldom gets that kind of distribution. And so what? Another story has already grabbed people's ever-roaming attention.

Attending a reception with three hundred people means exposing yourself to three hundred cameras that can send photos and videos with voice instantly into the cloud, complete with accompanying text and Photoshopping instructions. It gets even crazier. I have been followed into airport bathrooms by camera-carrying jerks looking for a money shot. I now use a closed stall.

Princess Diana was beloved, and she used her fame and position to advance many worthy causes. But her celebrity was a terrible cross to bear.

The challenge in public life is to keep your balance. Most people are decent, and want to reach out to you in kindness. Be pleasant to everyone who is pleasant and civil to you. Ignore the pests, hangers-on, and parasites. Always remember that celebrity is bestowed on you by the public; use the influence it gives you for worthwhile purposes and not just to pump up your ego. In other words, use your position for good, but don't let it go to your head. Don't believe all you hear or read about yourself, good or bad. Don't make your public life your full-time occupation, and hide frequently from the madding crowd.

Chapter Thirty-Nine
Speaking Is My Business

I have been a professional public speaker for most of my adult life. From my first day in my first unit as an Army officer, I had to speak to and teach troops. Over time I learned how to reach them, how to make the subject interesting, and how to persuade them that they had an interest in learning what I was teaching. Since they bored easily, a kit bag of attention-grabbing techniques was essential. You had to have a stable of jokes, and in those raw, male-only infantry days, the dirtier, the better.

In 1966, I was assigned to be an instructor at the Infantry School at Fort Benning. Before you're allowed in a classroom to teach two hundred officer students, you had to complete a several-week-long Instructor Training Course. There you learned thorough preparation of the subject material. You were taught eye contact, how not to cough, stammer, put your hands in your pocket, pick your nose, or scratch your itches. You were taught to stride across the stage, use a pointer, slides, and hand gestures, and how to raise and lower your voice to keep the students awake.

I graduated with honors and was turned loose to teach. But even after passing through the tough training course, I wasn't left on my own. The large classrooms in Infantry Hall had one-way glass windows in the back where your boss could watch your performance without your knowing it. Teaching there was hard work.

The most demanding class I taught was on filling out a Unit Readiness Report (where you measure the readiness of your troops by filling in blanks about the status of training, equipment, weapons, supplies, and so on). Nothing could be more boring. Worse, I had to teach it to officer candidates who had reached the end of the officer training program and were about to be commissioned. Most of them would be sent off to Vietnam. When they came to me, they had just returned from their graduation field exercise, three sleepless days in the Georgia pine woods, where it was either too hot or too cold. Mine was their last course, given at 4 p.m. on their last day. They came in from the field, took a hot shower to wash off the dirt, had a late spaghetti-and-meatball lunch, and then were turned over to me for fifty minutes in an air-conditioned classroom to learn how to fill out a readiness form they knew they wouldn't see again for years . . . and maybe ever.

The first couple of minutes always went okay. But then they started to drop. They tried to stay awake. They punched each other. Their tactical officers prowled the aisles, giving sharp stares and sharper pokes. Fifteen minutes in, the sound of heads hitting tables meant it was time for a good joke. At twenty minutes I warned them they could get killed in Vietnam if they didn't know how to fill out a readiness form. That bought me five minutes as they questioned whether or not I was nuts.

At about thirty minutes in, those who were totally lost to sleep were made to stand up and lean against the side walls. At forty minutes almost all of them were out of play. I had one gimmick left. I would pose a question and ask who wanted to answer it. Before they could duck that, I reached under the lectern and pulled out a very lifelike plucked rubber chicken. Swinging it over my head by the neck I launched it into the class. All two hundred pairs of eyes were wide open, watching the arc. Whoever it hit was directed to answer the question. The place broke up.

That gave me just the time I needed to summarize the class, congratulate them on the gold bars that would be pinned on them the next day, and wish them all the best as they headed overseas. Too many, sadly, did not return alive. To this day, I run into guys who say to me, "Hey, General, I'll never forget that damn chicken."

I also taught at Benning a class on amphibious warfare, alongside Marine Lieutenant Colonel P. X. Kelly (who went on to become the Commandant of the Marine Corps). Kelly was considered the best instructor at the school, and he taught me a lot. During the forty-five years since my instructor days, I've had fun telling Marines that I taught P.X. everything he knows about amphibious warfare.

What I learned in school and on the job at Benning stuck with me, and all I've done over the years is to build on that base. In my public life I have spoken to presidents and kings; I have spoken to large audiences and to small intimate groups. I have spoken at two Republican nominating conventions, and at too many congressional hearings to count.

When I left government in 1993, I embarked on a new career as a professional speaker, both here and abroad. Except for the four years when I was Secretary of State, public speaking has been my chief business activity and source of income. On my tax forms, I list my occupation as "Speaker"—or, if space permits, "Author/Speaker."

Although I had a number of other employment opportunities, I chose speaking over sitting on lots of corporate boards, working for a defense contractor, or taking a full-time job in academia or business. Since I have the freedom to decide how much speaking I want to do, I have the flexibility to take on less time-consuming business activities, engaging in nonprofit work, or just sitting around. At my age, the absolutely last thing I want is a full-time job that requires me to be at the same place every day, morning to night. No matter how exciting the work may be or how important the position, it's not for me anymore.

I love speaking for more reasons than time flexibility. For starters, it's great fun, and more important, it opens up new experiences and new learning. It allows me to plunge into worlds I never imagined. My audiences are businesses, trade associations, universities, or large motivational events. Every audience is different, and every one requires study. Who are they? What do they do? What's their purpose? What do they need from me? I have to orient my speech to them. I read annual reports, research organizations endlessly, and end up knowing enough about them to apply for a job.

I let clients know I can do whatever they want me to do; I can pitch content square or round.

It's always important to remember that a speaker has more than one responsibility. To begin with, he has a responsibility to his audience to give them what they need to hear. Sometimes it's what they expect; sometimes they may need to be shaken up. Second, if the speaker is on the podium representing an organization, he has a responsibility to the organization. He can't just go off on his own. When I was Secretary of State, I was speaking for the government of the United States. Most of my official State Department speeches were written, cleared, approved, and blessed. But I did extemporize from time to time. And finally, a speaker has a responsibility to himself. He owns whatever comes out of his mouth. He should never let himself speak words that he can't stand behind. When I spoke to the 1996 and 2000 Republican National Conventions, I wrote my own speeches. Nobody in the Republican National Committee told me what I was going to say. Of course I worked with an RNC representative and showed my speeches to the RNC the day before I gave them. They had no problems. But in both cases, the words were mine, not theirs.

I seldom use a text. But I have in my head lots of speech modules. I bring down for each audience the ones I need, modified when needed. I can pitch my speech at whatever level of sophistication the client wants.

Speech modules change over time. I drop or add elements to keep them fresh; or a new audience or a new need will require me to compose a new module.

Each speech follows a basic pattern. I start out talking about myself and what's going on in my life. Earlier in my speechmaking career, I told lots of jokes. I don't do that anymore. Instead, I tell self-deprecating stories about myself or my family, where Alma is often a central feature. My audiences don't expect casual, personal stories from a four-star general, and they welcome them. The stories warm up the audience and show a far more human person than the formal image I had to project when I was Chairman or Secretary of State. I open the door and let them into my real world for a little while.

In my speeches, I always include a segment on leadership. Drawing once again from my time at Fort Benning, I talk about mission, a sense of purpose, and the necessary connection between leadership and followership. I focus on taking care of the troops, on communicating selfless passion and intensity about the purpose of the organization, and on basic honor and honesty. Troops—followers—will only go up the hill for leaders who have character, integrity, and moral and physical courage.

Next, I broaden the presentation to talk about how the world has evolved and what forces are shaping the future. I then move into current events that are of interest to the audience. Though my ending changes constantly, I always end on a positive note. I want to leave audiences upbeat and encouraged.

I could relate a hundred anecdotes from my speaking experiences. Here are some of the most memorable—and that I learned the most from.

There was the evening in 2007 when I flew to Puerto Rico and drove to the Conquistador Hotel at the eastern end of the island. My client was the Bradford White water heater company, and I was to be a surprise speaker for a couple of hundred of the company's salespeople. Bradford White had been owned by an Australian holding company, but in 1992 the employees had bought the company and its Michigan factory from the foreign parent and had become independent. They are now proud to make all-American products whose high quality means they can successfully compete against anyone.

The company was run by Bob Carnevale, a street kid like me. We became instant friends. Just before going onstage, I asked him why he had dragged me all the way out here to be a surprise speaker.

"I didn't want any of my salespeople getting excited just to come here to meet and listen to you," he said. "I wanted them here so I could teach them how to sell more water heaters. You are just dessert."

I immediately understood why he was successful.

Many clients provide specific guidance about what they are looking for, but in all my years of speaking, one company stands out: Safelite AutoGlass, whose business is repairing car windshields. Tom Feeney, the president and CEO, was determined to increase the company's market share by showing their dealers how to get the best out of their people and their customer relations. For weeks they sent me notes, memos, and slides—eventually totaling an inch thick—about how they wanted me to talk about their leadership strategy at their Feed the Fire Leadership Meeting. I wish every company in America had their commitment to human resource development.

I didn't know much about the ups and downs of the housing market until 2007, when I addressed the International Housewares Association, whose members make knives, forks, plates, glasses, pots, and other housewares. Their sales, they explained, are a leading indicator of the housing market. If fewer knives, forks, and glasses are sold, then fewer new houses are being built. (Divorces and new bachelors will slightly alter those numbers.) Housewares manufacturers could tell me what's happening in housing before HUD, Fannie Mae, or Freddie Mac.

I'll never forget the 2007 Century 21 Real Estate convention in Las Vegas. It was overwhelming. Backstage were six interpretation booths—more interpreters than I often saw at the UN.

That's what it took to communicate with their worldwide corps of agents. Noticing a Chinese-language booth, I innocently asked my host if that was for Taiwan or the mainland. Both, I was told, and the greater presence was on the China mainland, with fifteen hundred sales offices and millions of property listings. "And, by the way," he added, "we are encouraging our agents to go back to wearing our famous Gold Jackets. We want to reinforce our culture. The Gold Jacket image is iconic and holds us together as a team."

Now that Century 21, Amway, Estée Lauder, and other consumer-oriented firms are penetrating China, the country will never be the same.

One of my happiest speaking engagements was at the 2011 Whataburger Family Convention in Dallas. Whataburger is a modestly sized, family-owned chain of burger restaurants, most of them located in the South, and now operated by the second generation of the Harmon Dobson family. Harmon started the chain half a century ago with a single store, and was determined to make the best burger possible. On day one a customer took a bite and exclaimed, "Whataburger!" The name stuck, and the chain grew to about seven hundred stores.

When I asked the current owners, three of Harmon's children, why they didn't have thousands of stores like the other burger chains launched about the same time, they answered, "We use the very freshest raw materials. We couldn't keep up our quality standards if we grew any larger. Nothing wrong with those other guys, but that's not our purpose." All the employees are referred to as family members, and they are all treated that way.

A third generation of teenagers is waiting to take over and maintain their grandparents' standards.

With some of my audiences, I'm shameless. In 1997, I was speaking to an American Trucking Association audience in Las Vegas, before heading to Salt Lake City for a youth event with Governor Mike Leavitt. During the Q&A I was asked what I would like to do in the next phase of my life. What a softball! I would love to be a trucker, I answered. A gentleman named Bill England jumped up and shouted, "That is going to happen today." Turns out his family owned C.R. England trucking company, with headquarters in West Valley, Utah.

I flew on to Salt Lake City and the youth rally. Afterward, the governor escorted me back to the airport. Waiting was a beautiful, fire-engine-red truck with a very long trailer attached. The nervous driver invited me to take the wheel. Governor Leavitt jumped in and got in the sleeping bunk behind the driver . . . probably because he knew he had never seen a really big accident. To the driver's evident relief, I managed to guide the rig around the airport grounds and back to the starting point without hitting anything or stripping any gears.

I might yet be a trucker in the next phase of my life.

I love this country. Everyplace I go to here gives me renewed energy. Every day and every client brings a new experience and a restorative dose of faith in America. Yes, we have troubles; we have always had troubles. But we have always overcome them. Traveling around the United States I see people hard at work, innovating, creating jobs, and believing they can succeed, just as they believe America will continue to succeed. They are good people, and as long as they are out there working away, I have no fear for our future.

Chapter Forty
On the Road

I spend a lot of time traveling, both here and overseas. On average, I am on the road two or three days a week, logging tens of thousands of miles a year. I usually travel alone; Alma seldom joins me. She has heard all my speeches and knows all too well the travel drill: arrive, sleep, give the speech or perform at the event, and leave. There will be no touring, shopping, or leisure. I try to minimize time away from home. For me, it is strictly work. Even so, I don't mind traveling. It opens up experiences that I wouldn't see and hear in Washington.

Of course, most of my time on the road is spent outside the event itself—in airports, planes, trains, limos, and hotels. I enjoy sitting in an airport gate area wearing a baseball cap, hiding behind wraparound sunglasses, and watching America go by. Yes, many of us need to go on a diet and get more exercise. And yes, a dress code would be a big plus. But people seem happy and busy. I love seeing young mothers wrestle with their little darlings and all the paraphernalia now required to sustain a kid. I love older folks increasingly able to manage smartphones and iPads. The growing number of wheelchairs waiting for planes shows how we are aging as a people, yet we're not just sitting around. I frequently drop in on the USO lounges to thank volunteers and chat with GIs. And I always watch with appreciation and admiration the mostly immigrant cleaning people who empty the trash, mop the floors, clean the latrines, and go about their work with quiet efficiency. It reminds me of my long-ago days mopping floors at the Pepsi-Cola bottling plant in Long Island City.

Nobody likes going through security, but I really can't complain about it. I was in the administration that set up the Transportation Security Administration. I stand in line and wait my turn like everyone else. Try to bump the line, and the Internet will make you an instant villain. I take what comes to me as gracefully as I can. But sometimes it's hard. Once at Reagan National Airport some sensor detected an explosive element on my hands. Examinations by two Explosive Ordnance Detachment teams and three supervisors finally got me sprung. It took thirty minutes. Pointing out that I had been Chairman of the JCS and Secretary of State did not do the trick. Afterward, they speculated that the alarm was caused by my morning blood pressure pill.

Short-distance air travel usually means getting crammed into a small Brazilian or Canadian plane. It's like flying in an MRI machine. The logo on the plane's tail may suggest a major airline, but it's always hard to tell who actually owns and flies it. Nevertheless, it gets you there, even if you need a chiropractor after you get pried out.

I have nothing but praise for crew members, flight attendants, gate agents, baggage handlers, porters, mechanics, and all the others who, under lots of pressure, keep us moving.

I go back and forth to New York regularly on the Acela, the closest train we have in this country to high speed. It is fast, comfortable, and dependable . . . and there's no TSA. I travel business class, but Alma always goes first class because of the service and the sandwich. (Grrr.) Many of my friends still fly the shuttle. But heaven help your schedule if there's bad weather somewhere over the East Coast, clotting up air travel from Maine to Key West.

On the ground, for the sake of efficiency and comfort, I always insist on a professional limo service and an ordinary sedan. I am too old to crawl into one of those stretch limos kids use for high school proms. I am not stuck up. I've just had too many experiences where a client, intent on chatting with me, will borrow a new car from a local dealer, and then, distracted and erratic, try to drive, talk, and figure out all the new knobs and switches.

I am not picky about hotels. Any will do, from a Days Inn to a Ritz-Carlton. But I avoid hotels where there's too much service. I don't need staff constantly bugging me to explain how to adjust the thermostat or turn down the bed. I don't need to rattle around large suites. When I sign in, I use an assumed name. Until writing this book I used Edward Felson, from, of course, one of my favorite movies, *The Hustler.*

My desires are mostly simple: Please give me a cheap clock radio; not one that needs printed instructions and plays my iPod. I am old; please make the numbers red and no less than three inches high. Get the cheapest one you can find, and tell people they are free to take it.

Give me a closet big enough to hang something in and not already filled up with a safe, iron, ironing board, and that silly folding suitcase rack, left over from the days of ocean liner suites.

Please, oh please, don't get fancy shower controls with handles that give you no clue how to turn it, push it, or pull it on and off. I only need one showerhead, not a decontamination sprinkler system. Put the Jacuzzi in the Honeymoon Suite.

I haven't really found a pressing need to have a television set or phone in the bathroom. Nor do I need a scale. And I'm really frightened by those padded and heated Japanese toilet seats in upscale hotels. The complex control panel suggests other things the toilet will do, but I have been afraid to try them and doubt the need.

Here's a biggie: please, please, put large print on the shampoo and conditioner tubes and bottles. Is it asking too much to let us know in a readable font that we're putting shampoo and not hand cream on our heads?

A simple coffeemaker, please. I don't need to grind coffee beans. This doesn't apply in Las Vegas, where they generally don't give you a coffee machine in your room. They want you downstairs pulling the slots while you wait your turn in the coffee shop line.

Keep the TV simple. I don't want to use it to go on the Internet or play games. Push a wrong button on the remote and you have to call room service to straighten it out.

Please cut the number of bolsters, cushions, and all the other stuff piled on beds that make it difficult to find pillows and have no functional purpose beyond encouraging female guests to do the same thing at home. Guys don't get this.

Lamp switches should be at the base of the lamp. Don't make me have to follow the wire down to a switch near the floor, or burn my hand feeling up toward the bulb.

Finally, we live in the information age. Please don't make us crawl under desks looking for a wall outlet for our iPhones, laptops, iPads, and other electronic gizmos that need feeding.

Otherwise, I enjoy traveling. I am always happy to be out where I can observe all the myriad varieties of Americans. And I love being on the speaking circuit, or in schools, Boys and Girls Clubs, charity events, and all the other wonderful activities going on around our nation. They keep us rolling forward.

Chapter Forty-One
Gifts

As you rise in rank in the Army, you pick up large numbers of plaques and certificates commemorating your various units and awards, and you accumulate large numbers of signed, framed photos from senior officers and other officials. These are displayed prominently on "me walls" in offices and home dens. After a few years, there's no need for paint or wallpaper; you've got enough plaques and photos to do the job.

By the time I became a colonel, I had quite a collection, more than any wall could hold. A charming older brigadier general, about to retire, frequently dropped in to my office. Because he was always a source of wise advice, I asked him what he was going to do with all his plaques when he retired.

"Colin, my wife and I have designed a beautiful log cabin in the Shenandoah Mountains. We plan to live there most of the time, enjoying the beauty of the mountains. And on cold winter nights, we will huddle on the couch in front of the fireplace, drink hot toddies, and throw the plaques in, one by one. Our kids won't want them."

Well, I ended up saving most of mine, now mostly housed in my archive collection at the National Defense University in Washington. Also at the archive and here at home are large collections of glass, Lucite, stone, and brass objets d'art. The most memorable of these is a dark slab of granite with my image and a dedication lasered into it. So help me, the thing looks like a pet's headstone. I am sure the folks who gave it to me had it made by a tombstone maker.

Military challenge coins are another popular gift, usually embossed with the crest and motto of the unit, and often with the name of the commander. Every 101st Airborne Division trooper was expected to carry a 101st Division challenge coin. Whenever or wherever in the world you met another trooper from the 101st, he would "challenge" you with his coin. If you didn't have yours, you had to buy him a drink and suffer deep embarrassment. I carried my 101st coin in my wallet for decades, until a little round spot on my bottom started to become ulcerous.

In the old days, challenge coins were given out sparingly, but sometime in the 1980s the tradition went viral. Many Army guys have dozens of them. Every unit and every senior leader has challenge coins; they spread them around to everyone they meet at any opportunity. Over time they have become more elaborate and more expensive, and more and more junior leaders and offbeat units have been passing them out. I've gotten personalized coins from a commissary officer, and even from a young sergeant who was a sedan driver. The practice has even spread to the civilian world. Cabinet officers and other civilian appointees give them out.

I started to push back when my coin collection went into the hundreds. It seemed like too much of an ego trip for the givers and a questionable use of funds (most, but not all, are paid for by the government). On the other hand, the troops love them and are eager to receive them, so the tradition has grown. I gave out challenge coins when I was Chairman and Secretary of State. I still have a small stash that I give out sparingly to, say, recovering GIs at Walter Reed Hospital, who seem to deeply appreciate them.

As I moved into more senior positions in government and traveled the world more frequently, gifts from foreign leaders started to pour in. Naturally, they placed a demand on me to respond. Congress constrains us not to spend more than about three hundred dollars on the gifts we give and, darn it, not to keep gifts worth more than three hundred dollars, as determined by appraisers in the department and the General Services Administration. My protocol office was creative within this limit in finding Americana gifts for our foreign guests and other visitors.

On one of his visits, my dear friend Igor Ivanov, foreign minister of the Russian Federation, gave me a bottle of vodka in the shape of an AK-47 assault rifle. Since someone in some office somewhere decided it was worth more than three hundred dollars, I couldn't keep it or drink it. Don't ask me how they figured that out. Sad to say, it is probably now stashed away in some government warehouse.

Clocks, watches, cufflinks, and pen sets have always been the gifts I like best to receive. I now have lots of clocks, watches, and pens, and I enjoy them all.

But then there are the portraits. Over the years, I have received several dozen portraits of me from various countries. We have a display of the better ones in our exercise room at home. It has always fascinated me that the way artists paint my face is a near-sure giveaway of where they came from. An artist cannot avoid adding his culture to your image. Thus, in a very excellent portrait by a famous Japanese painter, I bear a striking resemblance to Admiral Yamamoto. The one on Egyptian papyrus looks strikingly like Hosni Mubarak. The one from Romania kind of makes me into Dracula. The artist from the Detroit NAACP didn't think I looked black enough, so he broadened my nose and thickened my lips. The two paintings from Bermuda are both pastels, and oh so very mellow. Only thing missing is Jimmy Buffett playing "Margaritaville." I don't recall what we did with the one done in birdseed. Every time my staff moved it, they left behind a trail of birdseed.

Russian President Gorbachev once gave me a beautiful shotgun. Because I wanted to keep it, I paid my government $1,200 to buy it back from the American people.

After the breakup of the Soviet Union, I got lots of guns, bayonets, assault knives, and binoculars from leaders of the former Warsaw Pact countries. It was one way for them to unload their inventories and give gifts at no expense. Even my sharp-eyed appraisers couldn't pretend these things were worth more than three hundred dollars.

My French colleague Dominique de Villepin used to give me bottles of French red wine. He insisted red wine was the elixir of health and urged me never to drink white. For some strange reason, those bottles all broke before I could turn them in for appraisal.

Prime Minister Berlusconi of Italy loved to give American men gorgeous ties made by his favorite tailor and tie-maker. Too bad so many of them were stained and didn't make it to the appraiser. He once gave me a high-tech watch that doubled as an emergency homing device for pilots in the event of a crash. You pulled a wire antenna out of the side of the watch. I turned it in.

Aware of my service in Germany and my fondness for German beer in those old flip-top bottles with porcelain caps, Joschka Fischer, my German counterpart and the leader of the Green Party, brought me a case of fine German beer. On his next trip, I scratched my head to come up with a gift for him. Since he was the leader of the Green Party, I gave him a case of the empties to return for the deposit. But since he loved to cook out, I also gave him a set of barbecue tools.

President Nazarbayev of Kazakhstan, a very gracious host, gave a spectacular luncheon in his palace in the capital, Astana. Though the vodka toasts flowed freely, I managed to successfully defend our nation's honor. Well beforehand I had been alerted to one of his habits: if he liked a guest, he would take off his watch and give it to him. The guest was then expected to give the president his own watch in return. After lunch, we stumbled into a small elevator to head downstairs. There he took off his watch and presented it to me. I then took off my watch and proudly, with a hug, gave it to him. He got a Timex, I didn't.

Arab officials, especially from Gulf nations, are exceptionally generous. Their gifts are normally way, way over the three-hundred-dollar limit. They know we have to turn the gifts in, but they can do no less. It is a sign of their friendship and respect, and it's deeply embedded in their culture. The gifts were accepted in that spirit. I ended up with quite a collection of Arab daggers. Some were quite simple, and I kept them. Others, which were encrusted with jewels, were turned in.

One night in 2004, a very close Arab friend overheard Alma remark that her favorite car had been a 1995 Jaguar that I had long since sold. Shortly after I retired in 2005, an identical, completely restored 1995 Jaguar showed up in front of the house. Since I was no longer a government employee, I was legally able to keep it, and I did for a while, but regifted it just before the *Washington Post* got wind of it and wrote a story.

After leaving State, I continued to receive gifts from foreign governments. One Arab nation came very close to presenting me with a beautiful rug the week before I stepped down. But our sharp-eyed ambassador suggested to them that perhaps they should have it cleaned one more time and send to me after I retired. That lad will go far.

Finally, during my time as Chairman about twenty years ago, I was seated next to Arnold Schwarzenegger at a benefit dinner. "How do you stay in shape?" he asked.

"I jog," I told him, "but that's getting harder as I get older."

Several days later, a LifeCycle exercise bike showed up at the house. I used it for years, until more modern models came along. I still have Arnold's original bike in my basement. Since it is not something you can regift or easily dispose of, I'll let my kids figure out its destiny after I am gone.

Notwithstanding the fun I've had writing about exchanging them, we display in our home many wonderful gifts I've received over the years. Some are expensive, most not. They give us joy and fond memories of people and places all over the world we were privileged to visit and come to know well. And it gave us the opportunity to present to foreign friends gifts that convey our American spirit and tradition.

Chapter Forty-Two
Best and Worst

I am often asked what was my best or worst job, which of the presidents I worked for was best or worst, or who was my most important role model. Did I have a single best mentor? What was my greatest single achievement? What was my greatest single failure?

I don't answer those questions. To single out one success or one individual is to diminish many others that may be no less important. To single out your worst failure or least favorite person will surely make news . . . and your obituary writer's day.

I have a deeper reason for not answering. No matter how significant or life-changing your greatest hit or miss might be, neither even begins to define all of who you are. Each of us is a product of all our experiences and all our interactions with other people. To cite calculus, we are the area under the curve.

Being born into a good family might be the best thing that ever happened to me. But it was only a start. My parents were wonderful, but so was my parish priest, my accomplished older sister, my aunts and uncles, my teachers, my neighbors, and my buddies in the street. I was also shaped by the bullies in the street, the indifferent teachers, the people who looked at me and saw a colored kid who deserved to be treated as inferior on that basis alone.

The most influential people in my life will never show up on a Google search. They touched me long before anybody even dreamed of personal computers, the Internet, or search algorithms.

There was my first boss, the immigrant Russian Jewish toy store owner. "Finish your education, Colin," he told me. "Your future is not to work at my store."

There was Sammy Fiorino, whose shoe repair shop was just around the corner. Sammy taught me poker, how to get along with the neighborhood cops, and never to play poker with a man named Doc.

There was Miss Ryan, my high school English teacher, the only teacher whose name I can remember. She threatened and terrified me into working harder in her class than I'd ever worked at school. The English lessons she drilled into me were one of the greatest gifts I've ever received.

There was Colonel Harold C. Brookhart, my professor of military science at CCNY and a West Pointer. Colonel Brookhart nurtured me along through my early, less than polished attempts to do things the Army way. In my senior year, he sent me up to West Point for a three-day stay so I would be exposed to the principles and standards of the military caste and get a personal taste of my contemporaries. And in 1958, when I was about to set off for Fort Benning, he cautioned me with the best intentions not to expect Georgia to be like New York for a young black man.

There was Captain Miller in Germany, my first company commander. One day while out on maneuvers I realized my .45-caliber pistol was missing from my holster. Losing a weapon is serious business. I dutifully called him on the radio to report this horrible failure. When I got back to our camp site, he was waiting for me at the entrance. He handed me the pistol. "Local village kids found it," he told me, as a cold chill came over me. "They fired off one round. We heard it and got down there before they fired off another round and hurt someone. For God's sake, son, don't ever let that happen again."

He scared me to death. But when I checked the pistol I saw it hadn't been fired. In fact, it had never really been lost. Somebody had found it next to my cot, where it had fallen out of my holster as I had raced out of the camp. Miller had thought up a way to teach a promising kid lieutenant a lesson he'd never forget.

As a young black soldier I looked for inspiration to the few senior black officers then in the Army, and back in history to the black soldiers who had always served the nation proudly, even if the nation would not serve them. I had an obligation to stand on their shoulders and reach higher. I had to let my race be someone else's problem, never mine. I was an American soldier who was black, not a black American soldier.

Along the way there were many people I did not get along with and many who doubted my ability and potential. I learned from them to accept that they may be right and if so, to fix myself and keep moving on.

As I became more senior, more senior individuals entered my life, saw something in me worth using, reached out to guide me and move me along, and often pointed out weaknesses and problems. They all influenced me. I dare not begin to name them or this book would become much too long.

My focus here has been on early influences, because the shaping process begins in the early years. I often tell audiences that it begins the moment an infant hears her mother's voice and knows it is her mother's voice. It is that voice which speaks the language the infant will speak. It is that person who will make a remarkable bond of love with that child. It is that nurturing person who will begin imparting education, character, values, happiness, and kindness in the heart and mind of that child. In those early days, weeks, and months, the mother is the most important person whom that child knows. If she is not there or does not perform that role, the child has a much tougher road to travel.

I probably learned as much from failures and my naysayers as from my supporting rabbis. Failure comes with experience.

I recall a few years ago speaking at an elite and very highly structured Japanese high school. The kids were from good families and mostly very bright. After my remarks, designated kids from the honor roll lined up to ask me questions typed out on cards and fully vetted by their teachers.

After the first couple of questions, I turned away from the line and invited questions from anyone in the audience, with my eyes particularly focused on the back rows, where I used to try to sit.

One girl about thirteen years old raised her hand, and I called on her. "Are you ever afraid?" she asked.

"I am afraid every day," she continued. "I am afraid to fail." How brave she was to ask that question in public in a very structured Japanese high school.

Yes, I told her, I'm afraid of something every day, and I fail at something every day. Fear and failure are always present. Accept them as part of life and learn how to manage these realities. Be scared, but keep going. Being scared is usually transient. It will pass. If you fail, fix the causes and keep going.

The room was deadly silent. Every one of the young high achievers had the same question before their mind, even if they were too scared to put voice to it.

Failure is often solitary. Not so success. I am reminded of Michael Phelps, the swimmer who won a record eight gold medals at the 2008 Beijing Olympics. His physical ability and determination in the loneliness of a swimming lane are legendary. Yet he never fails to give credit to his parents, his coaches, his trainer, his team members, and all the others who helped him overcome attention deficit disorder and many other obstacles.

As successes come your way, remember that you didn't do it alone. It is always we.

Chapter Forty-Three
Hot Dogs

One of my favorite things to do is simply to walk along Park Avenue or Fifth Avenue in my hometown, New York, on a beautiful spring or fall afternoon. I love looking up at the classic buildings and churches and window gazing at the elite shops. Watching all the people go by is deeply moving; the whole world is represented, proving once again that we are a nation of nations.

Seeing all the varieties of people reminds me of the story of the Japanese billionaire who was asked by a Japanese TV interviewer which was his favorite city as he traveled around the world tending to his conglomerates. "New York," he said immediately.

"Why New York?" the interviewer asked him. "Why not Rome, Paris, London?"

"Because," he said, "New York is the only city in the world, where, when I walk down the street, people come up to me and ask for directions."

True, the whole world is there, and in so many other American cities.

On my walk, I always stop at the corner of a numbered cross street, where a Sabrett hot dog cart manned by an immigrant will always be stationed. I love those hot dogs, affectionately known to New Yorkers as "dirty water dogs" because they sit in a pot of near-boiling water.

I always must have one of them, adorned with mustard and that unique red onion relish I've only found in New York. It takes me back to my youth, when they only cost ten cents.

I even found time for it when I was Secretary of State. I would come out of my suite at the Waldorf-Astoria and stroll north up Park Avenue or perhaps over on Fifth Avenue. In those days I was surrounded by bodyguards, and there were usually a couple of New York City police cruisers rolling alongside to keep me from being whacked as I walked.

With my entourage I would walk up to the nearest hot dog peddler and order my hot dog. One poor guy, put off by the attention and all the police and guards, immediately stopped preparing my hot dog, thrust his hands up, and shouted, "I've got a green card, I've got a green card!" I assured him all was well and this was all about me, not him.

I still have to have a hot dog on my walk, but all the bodyguards and police cars are gone, as is the Waldorf suite. Shortly after leaving State, I went up to a hot dog stand on Fifth Avenue and ordered my standard fare. As the attendant was finishing up my hot dog, a look of recognition came across his face, but he struggled to pull up my name. "I know you," he said. "I see you on television." Then, as he handed me the hot dog, it hit him. "Ah, yes, of course, you're General Powell." I handed him the money, but he refused to take it. "No, General, no, you don't owe me anything. I've been paid. America has paid me. I will never forget where I came from, but now I am here, I am an American. I've been given a new life, and so have my children. Thank you, please enjoy the hot dog."

I thanked him and continued up the avenue, feeling a warm glow as the recognition came over me once again. What a country . . . still the same country that gave my immigrant parents that open door and welcome ninety years ago. We must never forget that has been our past; it is certainly our present and future.

There's a cute addition to this story. In 2009, I endorsed Mayor Mike Bloomberg for his third term as mayor of New York. His staff was looking for a photo op that would publicize the endorsement. They thought that a photo of the two of us at a restaurant would be a good idea. I suggested that a street corner purchase of hot dogs would be more New Yorkish and show the mayor in a more humble, "with the people" environment. They loved the idea, and the photo shoot was set up.

It was a cold morning, but I didn't have a coat on as I approached Mike on the corner. He was wearing an overcoat. Cameras started clicking and reporters and campaign staff were hovering. We walked up to the counter and I ordered two hot dogs. Mike interrupted and said to the guy, "I'll have my bun toasted." Ho, boy, this was not exactly a man-of-the-people request.

Nevertheless, it worked, and the photo was on page one, above the fold, of the *New York Times* the next morning.

I have even brought my love of hot dogs to the highest levels of diplomacy.

In April 2002, when he was still the vice president of the People's Republic of China, Hu Jintao visited Washington. While there, he was careful to keep his remarks very close to his government's positions. So, as we say in Washington, we pretty much exchanged standard official talking points.

One evening, I hosted the vice president at a State Department dinner where I wanted to do more than exchange position statements. Hu had just come from a visit to New York. I asked him about the visit. There were UN and other formal meetings, he told me, but not much else.

I told him he had visited New York, but hadn't seen it. The next time he visited I wanted to be his host. We would minimize the official events, and we and our wives would go to Broadway shows, walk along Forty-Second Street, and visit a variety of neighborhoods, to include Chinatown.

Above all, I told him, I wanted to buy him a hot dog from an immigrant peddler on a street corner. It took a while for the translator to figure it all out, but once he did, Hu broke into a smile. He thanked me and he told me he looked forward to it.

In November 2002, Hu became the president of the People's Republic of China. I have seen him several times over the intervening years, including a formal dinner in Washington after my retirement. He always spots me and has his aides escort me over to him. We shake hands and hug briefly. His first words, always in American English with a big smile, are "When do we get hot dogs?"

Hot dog diplomacy may not be earth-moving, but it allows two people to develop a human relationship that will help sustain an official relationship in good times and bad.

And remember, our country's opening to China began with a Ping-Pong match. I'm better at hot dogs.

Chapter Forty-Four
The Gift of a Good Start

During my time as Chairman of the Joint Chiefs of Staff, I often met senior foreign military leaders during my travels. Sometime during our initial meetings, I came to expect this question to come up: "When did you graduate from West Point?" Apparently they were still of the view that a West Point commission was the only way to get to the top.

"I didn't go to West Point," I replied, "as much as that would have been an honor."

"Well, did you attend the Citadel, the Virginia Military Institute, or Texas A&M?" they would then ask, referring to very well-known officer-producing institutions.

"No," I answered. "When I was entering college, a black person couldn't attend those colleges."

An embarrassed cough usually followed, and then came the next question: "Oh, well, where did you go?"

The answer was the City College of New York, in Harlem, not far from where I was born. I was commissioned through CCNY's ROTC program—the first ROTC graduate, the first black, and the youngest ever to become Chairman of the Joint Chiefs of Staff.

They immediately became curious. They had never heard of CCNY.

"It's a great school," I told them, "open to everyone." I'd usually go on to explain that CCNY was founded in 1847 and was then called the Free Academy. It was the first fully open, free college in America—a daring innovation in those days, as its president, a West Pointer, Dr. Horace Webster, declared on opening day in 1849:

"The experiment is to be tried, whether the children of the people, the children of the whole people, can be educated, and whether an institution of the highest grade, can be successfully controlled by the popular will, not by the privileged few."

The experiment succeeded. CCNY became a college of the first rank, but since it was free and drew from the immigrant and lower-income populations, it became known as the "Harvard of the Poor."

Time passes and I show up on campus in February 1954. I'm not sure how I got in. I was in no way an academic star. My high school grades were below the CCNY's admission standards. Was I given a preference? I don't know.

Earlier, when I was a teen looking at high schools, like most New York City kids I had dreamed of getting into the Bronx High School of Science, then the most prestigious high school in New York. (The story goes that Bronx Science has produced more Nobel Prize winners than France.) I didn't have a prayer.

Forty years later, I came across a devastating note from my junior high school guidance counselor: "Young Powell wants to attend the Bronx High School of Science. We recommend against it."

So, I went to Morris High School, where they had to let you in. I wasn't a bad student there, nor was I a great one, but I graduated and went on to CCNY.

At CCNY I was initially an engineering major, but quickly dropped it. Later I settled on geology, but by then I had discovered ROTC. I fell in love with ROTC, and with the Army.

After four-and-a-half no-cost, undistinguished academic years, the CCNY administration took pity on me and allowed my ROTC A grades to remain in my overall average. This brought my average up to a smidgen above 2.0, high enough to qualify for graduation. To the great relief of the faculty, I was passed off to the U.S. Army.

Nearly sixty years later, I am considered one of CCNY's greatest sons. I have received almost every award the school can hand out; an institute at CCNY has been named after me, the Colin L. Powell Center for Leadership and Service; and I have been titled a Founder and Distinguished Visiting Professor. Most of my professors have to be spinning in their graves over all that.

My city believed that kids like me deserved a shot at the top. The people of New York City were willing to be taxed to educate the "whole people"—poor kids like me with immigrant parents, Jews who couldn't get into other schools because they were Jews, young adults with jobs who could only go to night school (it might take them seven years to finish), kids who lived at home and came in every morning by subway or bus. Education like the one I got at CCNY was how the tired, poor, huddled masses yearning to breathe free were integrated into America's social and economic life. Education was—and still is—the Golden Door.

Though I only walked away with a diploma by the skin of my teeth, I did come out of college with a wonderful liberal arts education. I found in the years to come that I was able to perform well alongside my West Point, Citadel, VMI, and A&M buddies . . . as well as my buddies from other colleges and universities all over the country. We were all a band of brothers.

When I left the State Department in January 2005, time opened up for me to visit CCNY and see what the Powell Center had been doing since its founding eight years earlier. I sat in the college president's conference room and listened to each of about a dozen Powell Fellows tell me about themselves, what they were studying, and what they wanted to do with their lives. They were mostly minority, mostly immigrant, mostly first in their families to go to college, and mostly from low-income families. Many of them worked. But their eyes were bright, they were excited, they were hungry to do well. They had big ambitions, hopes, and dreams, and were working hard to succeed. Their words deeply moved me. They were just like I was more than fifty years earlier. CCNY was still the Harlem Harvard preparing another generation of public school winners. I told them how proud they made me and how I would be spending a lot of time at the Center.

In the years since that meeting, our programs have expanded enormously, with a focus on leadership training for these future leaders and on service learning so they could take their academic work into the community to help others. We changed the Center's name—originally the Center for Policy Studies—to the Center for Leadership and Service. The Center rapidly outgrew its two-room corner office. I hope that soon a Powell Center building will rise on the campus. It will not only house the Center, but also become a centerpiece for the entire campus and a gathering place for the people of central Harlem.

I'm proud that the Center has been named after me, but I'm no less proud that some seven new elementary and middle schools have also been named after me. I have adopted a school in Washington, D.C., and partnered it with the parishioners of my church in suburban Virginia. These mean more to me than any medal I have received. And additionally, as part of my passion for youth development, I served on the boards of trustees of Howard University and the United Negro College Fund and on the board of governors of the Boys and Girls Clubs of America.

I am frequently asked why youth programs and education have become a priority in my life. My answer is very simple: I want every kid to get the chance I had. No, West Point wasn't in the cards for me, but it showed me the standard I needed to attain. Morris High School and CCNY gave me the means to reach those standards.

I've learned a simple and obvious truth from my own education experience: We have to give every kid in America the access to public education that I received.

We need to place public education at the top of our priorities and the center of our national life.

Education has become my family's great crusade. In 1997, at the request of President Clinton and our other living former presidents, I founded the America's Promise Alliance to mobilize the country to give all our kids the basic skills and support they need to succeed in school and in life. Alma is now the chair of the Alliance, and our son Michael is on its board of directors.

America's Promise focuses on five basic promises we must make and deliver to our children. We promise them a responsible, caring loving adult in their lives to guide them along the right path. Where the family is unable to do that, we need to provide mentors. We promise them safe places in which to learn and grow, protected from the negative influences encountered in too many of our communities. We promise to try to provide every child with a healthy start and access to continuing health care. We promise our kids a good education with marketable skills. Finally, we promise them an opportunity to serve others so that they grow with the virtue of service embedded in their hearts. We have created a powerful alliance of partnerships with schools, nonprofit youth organizations, governments, and businesses to make sure we once again become a nation of graduates, not dropouts. We need to do this for the sake of the kids, for the sake of the future of the country we all love, and for the sake of our noblest ideals.

I love telling the story of my rocky education career to youngsters. My point is, it isn't where you start in life that counts, it is where you end up. So, believe in yourself, work hard, study hard, be your own role model, believe that anything is possible, and always do your best. Remember that your past is not necessarily your future.

Shortly after I retired from the Army in 1993, I was in West Palm Beach, Florida, giving a speech at the Kravis Center to a group of civic leaders raising money for the Boys and Girls Clubs of Palm Beach County. Before the event, I visited the Delray Beach Boys and Girls Club, also in West Palm Beach, a city where many of the less affluent and the workers who serve the affluent live. Maybe a hundred kids were sitting on the floor in front of me, ages ten to eighteen. I talked about growing up in Harlem and the Bronx and about my family and school experience. I tried to give them a Horatio Alger pitch. When I finished, I asked for questions. The little kids asked little-kid questions, like how much do you weigh, have you ever shot anyone, and what is your favorite color. The teens asked about my aspirations and my thoughts about running for president or vice president. Then a ten-year-old member raised his hand and asked, "I want to know if you think you would be where you are today if your parents didn't care whether you were alive or dead." He was talking about himself. My initial response was "I don't know." Then after a few seconds to gather my thoughts, I said, "You know, if your parents are not there for you, it doesn't mean the answers aren't there for you. The answers are here at the Boys and Girls Club and at your church and in your school. You come to this club every day. People are here waiting to help you, to teach you, to make sure you have fun. You can make it if you believe in yourself as much as they believe in you. I am not saying it will be easy, but the answers are there. You have to find them." I don't know if I convinced him, but I knew I had to do as much as I could to help him and others like him.

You can leave behind you a good reputation. But the only thing of momentous value we leave behind is the next generation, our kids—all our kids. We all need to work together to give them the gift of a good start in life.

Afterword
It's All About People

A couple of years ago I started jotting down stories, anecdotes, and experiences that had lodged firmly in my memory and for which I felt an affection. None of them was especially heavy or deep. None offered profound thoughts on the major issues of the day or on grand strategy. They were mostly human interest stories I thought I might use in speeches and public appearances. President Reagan used to keep a file of jokes. I had something similar in mind.

Most of my stories gave me a smile. For instance, one night Alma and I went to a movie in a local mall. As we headed back to our car, the lady parked next to us spotted us approaching. "Oh," she said, "I recognize you. You're . . . ?" She couldn't retrieve the name, and so I stood there for a moment to give her a chance to search her memory. Alma got into the car. After another long minute, I said to her, "Ma'am, I'm Colin Powell." She looked at me, bewildered, and said, "No, that's not it." She then got in her car and drove off. I'm often recognized as somebody who should be recognized, but I'm often mistaken for somebody else. Just the other day in the Atlanta Airport, a German tourist pointed me out to his wife. "That's General Schwarzkopf," he told her. When these mistakes happen, Alma won't allow me to immediately tell people I'm Denzel Washington. . . . If only we could choose who we're mistaken for.

Over time, the stories piled up, and I began to wonder if they might form the basis for a book. Most were similar to the stories I told in the first half of my memoir, *My American Journey*—personal stories about how I grew up, learned from good and bad experiences, and developed as an Army officer. People remember those stories far more than my coverage in the second half of the book of the serious and profound events of the 1980s and '90s—the end of the Cold War, Desert Storm, reorganization of the armed forces, the unification of Germany, and many more. Maybe historians will find interest in those pages, but seventeen years after *My American Journey* was first published, I am still asked about the personal stories, the stories about ordinary people. I have adapted many of them for this book.

When my pile of scribbled stories got to be sufficiently weighty, I showed them to a few close friends and trusted agents. Their response was gratifying. "These stories don't just make pleasant reading," I was told. "They show you learning something important about life and leadership. Other people may also learn from them. Why don't you turn them into a book?"

Most of the chapters in the book that has emerged from my pile of scribbled stories are about people I have encountered in my life—family, friends, colleagues, bosses, followers, adversaries, an enemy or two, some rich, some poor, some high and mighty, some not so high and mighty.

I have learned from most of the people I've met, and I have tried to inspire the people I have led. Life and leadership can't be about *me*. They have to be about *us*. They have to be about people. I remember attending a small office promotion ceremony in Washington in the early 1970s. I can't remember who was being promoted or where the ceremony was held. But I vividly remember that Admiral Hyman G. Rickover, the father of the nuclear Navy, spoke there. Rickover was as crusty and demanding a leader as anyone has ever seen, with enormous pull on Capitol Hill.

After the promotion ritual, Rickover was asked to say a few words. His words have stayed with me:

"Organizations don't get things done. Plans and programs don't get things done. Only people get things done. Organizations, plans, and programs either help or hinder people."

The wisdom of his words has shaped my life.

Back in 1972 I was a White House Fellow. Ever since I've felt close to the Fellows; every year I speak to their incoming class; and every year I make this point to them: No good idea succeeds simply because it is a good idea. Good ideas must have champions—people willing to believe in them, push for them, fight for them, gain adherents and other champions, and press until they succeed. I follow up with a related truth: Bad ideas don't die simply because they are intrinsically bad. You need people who will stand up and fight them, put themselves at risk, point out the weaknesses, and drive stakes through their hearts.

A life is about its events; it's about challenges met and overcome—or not; it's about successes and failures. But more than all of these put together, it's about how we touch and are touched by the people we meet. It's all about the people. I hope that comes through clearly in the pages you have just read.

The people in my life made me what I am.

Acknowledgments

I thank my wife of almost fifty years, Alma, for her support and understanding as I worked on the book. Her quiet and steadfast encouragement made all the difference as I labored away in my home office, fondly known as the bunker. I couldn't have done it without her; or without the love and needling of our children, Michael, Linda, and Annemarie.

My agent, Marvin Josephson, did yeoman service again in working with HarperCollins to publish the book. Marvin was my agent for *My American Journey*, close to twenty years ago. During the years between the two books, he served as a dear friend and close confidant. He is the best at what he does.

Tony Koltz, a distinguished author and collaborator, was my right-hand partner and indispensable colleague in bringing order and precision to the book. I am deeply appreciative of his hard work and dedication to the project. When I engaged Tony, I didn't know I would also pick up the assistance of his wife, Toni Burbank. Toni is a well-known and highly respected editor who served as the tiebreaker when Tony and I got into a disagreement. She usually voted for Tony, and both of them were usually right. I occasionally overruled them just to prove Generals are in charge, or think they are.

Peggy Cifrino and Leslie Lautenslager have been my beloved assistants for going on twenty years, to include my time in the State Department. Peggy runs my office and my life. Leslie works at the Washington Speakers Bureau, and I am her account. She moves me around the world with efficiency and courtesy. We have been a great team for a long time. They both have worked on the book from the very beginning, offering ideas, commentary, and criticism. A lot of this book draws from the way they go about their jobs. They are treasures.

My thanks to Edwin Lautenslager, Leslie's brother, who organized the first rough draft of the book, which gave me something to show people. While he got me started, it was Margaret Lautenslager, his and Leslie's mom, who did the final read with the discerning eye of a demanding schoolteacher grading a pupil.

I can't say enough about the team at HarperCollins. They immediately saw merit in the book, and I am deeply appreciative of all they have done to shape it, promote it, and bring it to the public. Special thanks to Tim Duggan, my editor, for his thoroughness and all his suggestions. His most important contribution was understanding what I was trying to do and helping to make the vision a reality. My thanks to the entire HarperCollins team includes Brian Murray, CEO; Jonathan Burnham, publisher; Kathy Schneider, associate publisher; Tina Andreadis, publicity director; Beth Harper, publicist; and Emily Cunningham, assistant editor; and all who work for them.

Susan Lemke and her staff, who manage my papers at the National Defense University, provided superb assistance in locating obscure papers and ancient photos at the drop of a hat.

I received invaluable guidance from a close group of friends who read the manuscript as it developed. The most anticipated comment was from my former collaborator, Joe Persico, who worked with me on my memoir, *My American Journey.* I was greatly relieved when Joe complimented me and Tony on the manuscript.

Marybel Batjer, one of my closest friends and colleagues, provided excellent suggestions. Tina Brown and Harry Evans gave me early encouragement and perceptive observations for which I am grateful. Grant Green and Bruce Morrison were very helpful with the chapter on Brainware. Larry Wilkerson helped with the UN chapter that he lived through with me.

I also benefitted from so many others, too numerous to mention. This is especially the case with the GIs I served with who provide most of the inspiration for this book. I thank them all.

—Colin Powell

About the Author

COLIN POWELL was born in New York City in 1937. He is a retired four-star general in the United States Army and has earned numerous military and civilian honors. He has served four presidential administrations in a variety of roles, most recently as Secretary of State from 2001 to 2005. He lives in Virginia.

TONY KOLTZ has co-authored with Tom Clancy the memoirs of Generals Fred Franks, Chuck Horner, Carl Stiner, and Anthony Zinni. He has also co-authored two additional books with General Zinni. He lives in New York City.

